Finite Element Analysis
and Applications

Finite Element Analysis and Applications

R. Wait

Department of Statistics and Computational Mathematics,
University of Liverpool

and

A. R. Mitchell

Department of Mathematical Sciences
University of Dundee

WITHDRAWN

A Wiley–Interscience Publication

JOHN WILEY & SONS
Chichester · New York · Brisbane · Toronto · Singapore

Library of Congress Cataloging in Publication Data:

Wait, R.
  Finite element analysis and applications.
  'A Wiley–Interscience publication.'
  Bibliography: p.
  Includes index.
  1. Finite element method. I. Mitchell, A. R.
II. Title.
TA347.F5W35 1985   620′.001′515353   84–25828

ISBN 0 471 90677 8 (cloth)
ISBN 0 471 90678 6 (paper)

British Library Cataloguing in Publication Data:

Wait, R.
  Finite element analysis and applications.
  1. Finite element method
  I. Title   II. Mitchell, A. R.
  515.3′53   TA347.F5

ISBN 0 471 90677 8 (cloth)
ISBN 0 471 90678 6 (paper)

Printed and bound in Great Britain

To
Fiona and Ann

Preface

The finite element method for the solution of problems in engineering and science has been the subject of considerable effort over the last three decades. The origins of the modern form of the method, as a mathematical technique for the numerical solution of partial differential equations, can be traced back to Courant in 1943. The underlying theory is even older and it is associated with the names of the turn-of-century mathematicians Ritz and Galerkin. The true power of the method was revealed by the development of the method by structural engineers using the early digital computers available in the 1950s. The rigorous mathematical analysis followed at a much later date and is often viewed as superfluous.

The methods for time-dependent problems, particularly those for hyperbolic equations, have been studied for a much shorter period and the development of both the practical methods and the theoretical analysis is less advanced than for the static equilibrium problems on which the method was raised.

In this book, the authors have attempted to provide an introduction to the method by considering both the theory and the practice. The book is aimed at final-year undergraduate and first-year postgraduate students in mathematical sciences or engineering. No specialized mathematical knowledge beyond a familiarity with calculus and elementary differential equations is assumed. The applications are drawn from many areas and no knowledge of structural mechanics, or any other branch of engineering science, is assumed. The abstract mathematics is kept to a minimum and is concentrated in a single chapter. An earlier version of this text was published by Wiley in 1977 as *The finite element method in partial differential equations*.

Chapter 1 provides an introduction to the concept of piecewise polynomial approximation, followed by a review of the elementary abstract analysis needed for Chapters 4 and 5. Chapter 2 is a self-contained description of the variational principles that form the basis of the Ritz formulation of finite element methods. The various forms of basis functions (or shape functions) used in the most common finite element approximations are detailed in Chapter 3, while Chapter 4 provides details of the methods. After a brief outline of Ritz methods, the majority of Chapter 4 describes Galerkin methods based on the weak solutions introduced in Chapter 1. Chapters 3 and 4 can be read more-or-less in parallel.

The discussion of the methods of approximation is continued in Chapter 5 to include time-dependent problems, while Chapter 6 contains the mathematical error analysis of finite element approximations. Chapter 7 provides a few example of applications that illustrate areas of study in 1984.

Much of this material has been presented in the form of honours and MSc lectures at the Universities of Liverpool and Dundee. The authors acknowledge the advice and assistance of colleagues and former students too numerous to mention individually. Finally thanks are due to Doris Barton and Ros Hume for the typing of the manuscript.

Liverpool and Dundee   **R. Wait**
1985   **A. R. Mitchell**

Contents

1

Introduction

In the numerical solution of partial differential equations by finite element methods, the differential equation in terms of a continuous unknown solution is replaced by a system of algebraic equations in terms of the parameters defining the approximate solution. The essence of the finite element method is to partition the domain of the problem into non-overlapping *elements* and to provide an approximate solution that has a simple form within each element. The local representations are then patched together to form a global solution of the desired smoothness. As the local form of the solution is to be kept simple, accuracy is achieved by making the elements as small as possible; this is turn means that the approximate problem is defined by means of a large number of equations. A piecewise polynomial function defined in terms of the values at nodes, defined by the element geometry, leads to a *sparse* coefficient matrix (i.e. few nonzeros per row) and to the possibility of solving bigger problems.

1.1 Approximation by Piecewise Polynomials

Consider initially the problem of approximating a real-valued function $f(x)$ over a finite interval of the x-axis. A simple approach is to break up the interval into a number of non-overlapping subintervals and to interpolate linearly between the values of $f(x)$ at the endpoints of each subinterval (see Figure 1.1). If there are n subintervals denoted by $[x_i, x_{i+1}](i = 0, 1, 2, \ldots, n-1)$, then the piecewise linear approximating function depends only on the function values $f_i(= f(x_i))$ at the nodal points $x_i(i = 0, 1, 2, \ldots, n)$. In a problem where $f(x)$ is given implicitly by an equation (differential, integral, functional, etc.), the values f_i are the unknown parameters of the problem. In the problem of interpolation, the values f_i are known in advance.

In the subinterval $[x_i, x_{i+1}]$, the appropriate part of the linear approximating

1

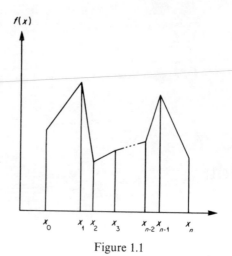

Figure 1.1

function is given by

$$p_1^{(i)}(x) = \alpha_i(x)f_i + \beta_{i+1}(x)f_{i+1} \quad (x_i \leqslant x \leqslant x_{i+1}) \tag{1.1}$$

where

$$\alpha_i(x) = \frac{x_{i+1} - x}{x_{i+1} - x_i} \quad \text{and} \quad \beta_{i+1}(x) = \frac{x - x_i}{x_{i+1} - x_i} \quad (i = 0, 1, 2, \ldots, n-1).$$

The local functions $\alpha_i(x)$ and $\beta_i(x)$ are known as *shape functions* and are only defined within an individual element. The values f_i and f_{i+1} are *nodal parameters*. In this example there are two parameters per element and hence the elements are said to have two *degrees of freedom*.

Hence the piecewise approximating function over the interval $x_0 \leqslant x \leqslant x_n$ is given by

$$p_1(x) = \sum_{i=0}^{n} \varphi_i(x)f_i \tag{1.2}$$

where

$$\varphi_0(x) = \begin{cases} \dfrac{x_1 - x}{x_1 - x_0} & (x_0 \leqslant x \leqslant x_1) \\ 0 & (x_1 \leqslant x \leqslant x_n) \end{cases}$$

$$\varphi_i(x) = \begin{cases} 0 & (x_0 \leqslant x \leqslant x_{i-1}) \\ \dfrac{x - x_{i-1}}{x_i - x_{i-1}} & (x_{i-1} \leqslant x \leqslant x_i) \\ \dfrac{x_{i+1} - x}{x_{i+1} - x_i} & (x_i \leqslant x \leqslant x_{i+1}) \\ 0 & (x_{i+1} \leqslant x \leqslant x_n) \end{cases} \tag{1.3}$$

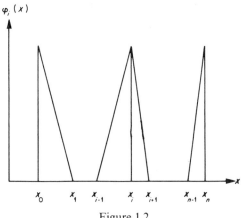

Figure 1.2

and

$$\varphi_n(x) = \begin{cases} 0 & (x_0 \leqslant x \leqslant x_{n-1}) \\ \dfrac{x - x_{n-1}}{x_n - x_{n-1}} & (x_{n-1} \leqslant x \leqslant x_n) \end{cases}$$

are pyramid functions illustrated in Figure 1.2. The pyramid functions given by (1.3) represent an elementary type of basis function. In particular the basis functions $\varphi_i(x)(i = 1, 2, \ldots, n-1)$ are identically zero except for the range $x_{i-1} \leqslant x \leqslant x_{i+1}$, and are said to have local support.

The shape functions can be identified as restrictions of the basis functions to a single element. If φ_i^e denotes the restriction of φ_i to element $e = [x_i, x_{i+1}]$ then $\alpha_i = \varphi_i^e$ and $\beta_{i+1} = \varphi_{i+1}^e$.

A useful procedure in constructing basis functions which involve function values only is to introduce the standard coordinate

$$X = \frac{x}{h} - i \quad i = 0, 1, 2, \ldots, n$$

where we assume a uniform element of size h. In this new coordinate, (1.3) simplifies to give

$$\varphi_i(X) = \begin{cases} 0 & X \leqslant -1 \\ 1 + X & -1 \leqslant X \leqslant 0 \\ 1 - X & 0 \leqslant X \leqslant +1 \\ 0 & X \geqslant +1 \end{cases}$$

basis functions which are usually called piecewise linear 'hat' functions, and are illustrated in Figure 1.3.

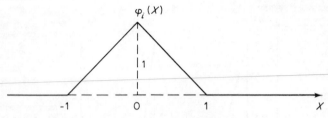

Figure 1.3 Piecewise linear function $\varphi_i(X)$

If we now approximate the function $f(x)$ by a piecewise *quadratic* function, the latter depends on the function values f_i at the nodal points x_i $(i = 0, 1, 2, \ldots, n)$ (integer points) and the function values at intermediate points, usually chosen to be $x_{j+1/2}(j = 0, \ldots, n - 1)$ (half-integer points). This time the piecewise approximating function over the range $x_0 \leqslant x \leqslant x_n$ is given by

$$p_2(x) = \sum_{i=0}^{n} \psi_i(x)f_i + \sum_{j=0}^{n-1} \chi_{j+1/2}(x)f_{j+1/2}$$

where the basis functions, in terms of the standard coordinates, are

$$\psi_i(X) = \begin{cases} 0 & X \leqslant -1 \\ 1 + 3X + 2X^2 & -1 \leqslant X \leqslant 0 \\ 1 - 3X + 2X^2 & 0 \leqslant X \leqslant +1 \\ 0 & X \geqslant +1 \end{cases}$$

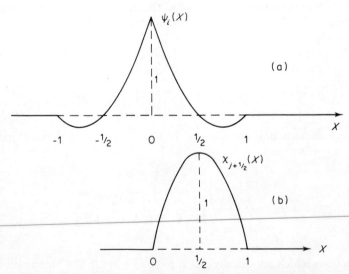

Figure 1.4 Piecewise quadratic functions: (a) $\psi_i(X)$; (b) $\chi_{j+\frac{1}{2}}(X)$

with $X = x/h - i$ at the integer points, and

$$\chi_{j+1/2}(X) = \begin{cases} 0 & X \leqslant 0 \\ 4X - 4X^2 & 0 \leqslant X \leqslant +1 \\ 0 & X \geqslant +1 \end{cases}$$

with $X = x/h - j$ at the half-integer points. These basis functions are illustrated in Figure 1.4.

In general, the first derivatives of the piecewise approximating polynomials $p_1(x)$ and $p_2(x)$ are not the same as $f(x)$ even at the nodes. Consequently we now look at the possibility of constructing an approximating function which has the same values of function and first derivative as $f(x)$ at the nodal points $x_i (i = 0, 1, 2, \ldots, n)$. In mathematical terms, we have to construct a piecewise cubic polynomial $p_3(x)$ such that

$$D^k f(x_i) = D^k p_3(x_i) \quad (k = 0, 1; i = 0, 1, 2, \ldots, n)$$

where $D = d/dx$. In the subinterval $[x_i, x_{i+1}]$, the appropriate part of the approximating cubic polynomials is given by

$$p_3^{(i)}(x) = \alpha_i(x) f_i + \beta_{i+1}(x) f_{i+1} + \gamma_i(x) f_i' + \delta_{i+1}(x) f_{i+1}' \tag{1.4}$$

where

$$\alpha_i(x) = \frac{(x_{i+1} - x)^2 [(x_{i+1} - x_i) + 2(x - x_i)]}{(x_{i+1} - x_i)^3}$$

$$\beta_{i+1}(x) = \frac{(x - x_i)^2 [(x_{i+1} - x_i) + 2(x_{i+1} - x)]}{(x_{i+1} - x_i)^3}$$

$$\gamma_i(x) = \frac{(x - x_i)(x_{i+1} - x)^2}{(x_{i+1} - x_i)^2} \tag{1.5}$$

and

$$\delta_{i+1}(x) = \frac{(x - x_i)^2 (x - x_{i+1})}{(x_{i+1} - x_i)^2}$$

$(i = 0, 1, 2, \ldots, n - 1)$ and where $'$ denotes differentiation with respect to x. The piecewise approximating function over the interval $x_0 \leqslant x \leqslant x_n$ is given by

$$p_3(x) = \sum_{i=0}^{n} [\varphi_i^{(0)}(x) f_i + \varphi_i^{(1)}(x) f_i'] \tag{1.6}$$

where the piecewise cubic functions $\varphi_i^{(0)}(x), \varphi_i^{(1)}(x)(i = 0, 1, 2, \ldots, n)$ are easily obtained from (1.5). The basis functions $\varphi_i^{(0)}(x)$ and $\varphi_i^{(1)}(x)(i = 1, 2, \ldots, n - 1)$ are illustrated in Figure 1.5.

The basis functions in (1.2) and (1.6) arise from particular cases of *piecewise Hermite interpolation* (or approximation) for a partitioned interval. In more

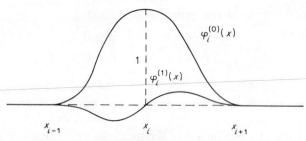

Figure 1.5 Piecewise Hermite cubic functions

general terms, let $\Pi : a = x_0 < x_1 < \ldots < x_n = b$ denote any partition of the interval $R = [a, b]$ on the x-axis. For a positive integer m, and a partition Π of the interval, let $H = H^{(m)}(\Pi, R)$ be the set of all real-valued piecewise polynomial functions $w(x)$ defined on R such that $w(x) \in C^{m-1}(R)$ and $w(x)$ is a polynomial of degree $2m - 1$ on each subinterval $[x_i, x_{i+1}]$ of R. Given any real-valued function $f(x) \in C^{m-1}(R)$, then its unique piecewise Hermite interpolate is the element $p_{2m-1}(x) \in H$ such that

$$D^k f(x_i) = D^k p_{2m-1}(x_i) \quad \begin{cases} (0 \leqslant k \leqslant m-1) \\ (0 \leqslant i \leqslant n). \end{cases} \tag{1.7}$$

The particular cases $m = 1, 2$ have already been dealt with and produce the basis functions given by (1.2) and (1.6) respectively. Error estimates for piecewise Hermite interpolates are given by Birkhoff *et al.*, (1968).

In problems where only $f(x)$ has to be determined, it is often undesirable to introduce derivatives of $f(x)$ as additional parameters and so cause a considerable increase in the order of the system of equations to be solved. Consequently a very desirable property in piecewise functions might be continuity of derivatives at the points at which pieces of the polynomials meet without introducing the values of the derivatives as additional unknown parameters.

Such smooth interpolation functions are known as splines. A spline approximation can be written in terms of smooth piecewise polynomial basis functions known as B-splines. With

$$B_0(X) = \begin{cases} 1 & -\tfrac{1}{2} < X < \tfrac{1}{2} \\ 0 & |X| \geqslant \tfrac{1}{2} \end{cases} \tag{1.8}$$

it is possible to define the higher order splines in terms of *convolutions*

$$B_i(X) = \int_{-\infty}^{\infty} B_{i-1}(X - Y) B_0(Y) \, \mathrm{d}Y \quad (i = 1, 2, \ldots). \tag{1.9}$$

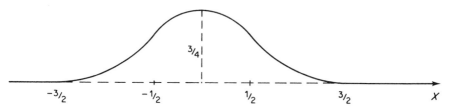

Figure 1.6 Piecewise quadratic spline $B_2(X)$

Clearly from the definition of $B_0(Y)$

$$B_1(X) = \int_{-1/2}^{1/2} B_0(X - Y)\, dY.$$

If $|X| \geqslant 1$ then since $Y \in (-\frac{1}{2}, \frac{1}{2})$ it follows that $|X - Y| \geqslant \frac{1}{2}$ and hence $B_0(X - Y) = 0$ and so

$$B_1(X) = 0 \quad |X| \geqslant 1.$$

For $|X| < 1$ consider $X > 0$ and $X \leqslant 0$ separately. If $X > 0$, then

$$\int_{-1/2}^{1/2} B_0(X - Y)\, dY = \int_{X-1/2}^{1/2} 1 \, dY = 1 - X$$

while for $X \leqslant 0$,

$$\int_{-1/2}^{X+1/2} 1 \, dY = X + 1.$$

Thus

$$B_1(X) = \begin{cases} 0 & X \leqslant -1 \\ 1 + X & -1 \leqslant X \leqslant 0 \\ 1 - X & 0 \leqslant X \leqslant 1 \\ 0 & X \geqslant 1. \end{cases}$$

This is the piecewise linear hat function illustrated in Figure 1.3. It is only by using the piecewise quadratic $B_2(X)$ (Figure 1.6) that it is possible to construct an approximation with continuous derivatives.

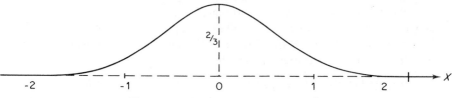

Figure 1.7 Piecewise cubic spline $B_3(X)$

The piecewise cubic $B_3(X)$ (Figure 1.7) has continuous second derivatives. A cubic spline approximation can then be written as

$$p_3(x) = \sum_i \varphi_i(x)c_i \tag{1.10}$$

where

$$\varphi_i(x) = B_3\left(\frac{x}{h} - i\right). \tag{1.11}$$

Note that now

$$\varphi_i(x_j) = 0 \quad \text{iff} \ |i - j| > 2$$

and

$$c_i \neq f_i \equiv f(x_i). \tag{1.12}$$

The convolution formula is a convenient mathematical definition for B-splines and we shall return to it in the discussion of spline collocation in Chapter 4. The most stable method of evaluating B-splines and hence splines in general is to use the recurrence formulae derived by Cox (1972) and de Boor (1972).

Exercise 1.1 Verify that the convolution is associative and that in particular

$$B_{2q+1}(X) = \int B_{2q}(Y)B_0(X - Y)\,\mathrm{d}Y$$

$$= \int B_q(Y)B_q(X - Y)\,\mathrm{d}Y.$$

1.1.1 Bivariate approximation

We now consider the problem of approximating a real-valued function of two variables by piecewise continuous functions over a bounded region R with boundary ∂R. The region is divided up into a number of elements and the particular shapes of region considered at this stage are (1) rectangular and (2) polygonal.

(1) *Rectangular region*   The sides of this region are parallel to the x- and y-axes, and the region is subdivided into similar rectangular elements by drawing lines parallel to the axes. Let the rectangular region be $[x_0, x_m] \times [y_0, y_n]$ and a typical element be $[x_i, x_{i+1}] \times [y_j, y_{j+1}]$, where $x_{i+1} - x_i = h_1$ and $y_{j+1} - y_j = h_2$ $(0 \leqslant i \leqslant m-1, 0 \leqslant j \leqslant n-1)$ (see Figure 1.8). The bilinear form which interpolates $f(x, y)$ over the rectangular element is

$$p_1^{(i,j)}(x, y) = \alpha_{i,j}(x, y)f_{i,j} + \beta_{i+1,j}(x, y)f_{i+1,j} + \gamma_{i,j+1}(x, y)f_{i,j+1}$$
$$+ \delta_{i+1,j+1}(x, y)f_{i+1,j+1} \tag{1.13}$$

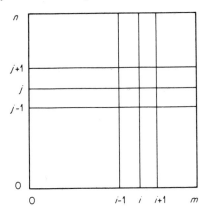

Figure 1.8

where

$$\alpha_{i,j}(x, y) = \frac{1}{h_1 h_2}(x_{i+1} - x)(y_{j+1} - y)$$

$$\beta_{i+1,j}(x, y) = \frac{1}{h_1 h_2}(x - x_i)(y_{j+1} - y)$$

$$\gamma_{i,j+1}(x, y) = \frac{1}{h_1 h_2}(x_{i+1} - x)(y - y_j)$$

and

$$\delta_{i+1,j+1}(x, y) = \frac{1}{h_1 h_2}(x - x_i)(y - y_j)$$

$(0 \leqslant i \leqslant m - 1; 0 \leqslant j \leqslant n - 1)$. The piecewise approximating function over the region $[x_0, x_m] \times [y_0, y_n]$ is given by

$$p_1(x, y) = \sum_{i=0}^{m} \sum_{j=0}^{n} \varphi_{i,j}(x, y) f_{i,j}. \qquad (1.14)$$

The basis functions $\varphi_{i,j}(x, y)(1 \leqslant i \leqslant m - 1; 1 \leqslant j \leqslant n - 1)$ are identically zero except for the rectangular region $[x_{i-1}, x_{i+1}] \times [y_{j-1}, y_{j+1}]$, and so have local support (see Exercise 1.4 and Figure 1.8).

The case just considered is the simplest example of piecewise *bivariate* Hermite interpolation (or approximation) over a rectangular region subdivided into rectangular elements. In more general terms, for any positive integer l, and any subdivision of the rectangle R into rectangular elements, let $H = H^{(l)}(R)$ be the collection of all real-valued piecewise polynomials $g(x, y)$ defined on R such that $g(x, y) \in C^{l-1,l-1}(R)$ and $g(x, y)$ is a polynomial of degree $2l - 1$ in each variable x

and y on each rectangular element $[x_i, x_{i+1}] \times [y_j, y_{j+1}](0 \leqslant i \leqslant m-1; 0 \leqslant j \leqslant n-1)$ of R. Given any real-valued function $f(x, y) \in C^{l-1, l-1}(R)$, then its unique piecewise Hermite interpolant is the element $p_{2l-1}(x, y) \in H$ such that

$$D^{(p,q)} f(x_i, y_j) = D^{(p,q)} p_{2l-1}(x_i, y_j)$$

for all $0 \leqslant p, q \leqslant l-1, 0 \leqslant i \leqslant m-1, 0 \leqslant j \leqslant n-1$. The particular case $l = 1$ has already been dealt with and leads to bilinear basis functions of the type shown in Exercise 1.4. The case $l = 2$ is covered in Exercise 1.5. The interested reader is again referred to Birkhoff *et al.* (1968) for error estimates of bivariate Hermite interpolation.

(2) *Polygonal region*   This can either be a region in its own right or an approximation to a region of any shape. The polygon is subdivided in an arbitrary manner into triangular elements. In a typical triangular element with vertices $(x_i, y_i)(i = 1, 2, 3)$ (see Figure 1.9) the linear form which interpolates $f(x, y)$ over the triangular element is

$$p_1(x, y) = \sum_{i=1}^{3} \alpha_i(x, y) f_i$$

where $f_i = f(x_i, y_i)(i = 1, 2, 3)$. The coefficients $\alpha_i(x, y)(i = 1, 2, 3)$ are given by

$$\alpha_1(x, y) = \frac{1}{C_{123}} (\tau_{23} + \eta_{23} x - \xi_{23} y)$$

$$\alpha_2(x, y) = \frac{1}{C_{123}} (\tau_{31} + \eta_{31} x - \xi_{31} y) \tag{1.15}$$

and

$$\alpha_3(x, y) = \frac{1}{C_{123}} (\tau_{12} + \eta_{12} x - \xi_{12} y)$$

where $|C_{123}|$ is twice the area of the triangle, and

$$\tau_{ij} = x_i y_j - x_j y_i$$
$$\xi_{ij} = x_i - x_j \quad (i, j = 1, 2, 3)$$

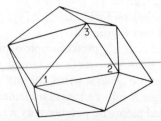

Figure 1.9

and

$$\eta_{ij} = y_i - y_j.$$

The functions given by (1.15) are of course only parts of the complete basis functions associated with vertices of a triangular network. The complete basis function with respect to any vertex is obtained by summing the appropriate parts associated with the triangles adjacent to the vertex. For example, the vertex 1 in Figure 1.9 has five adjacent triangles and so the basis function associated with this vertex has five parts. The complete basis function is known as a pyramid function.

Exercise 1.2 Show that the cubic polynomial $p_3(x)$ which takes the values

$$p_3(0) = f_0 \quad p_3(1) = f_1 \quad p_3'(0) = f_0' \quad p_3'(1) = f_1'$$

is given by

$$p_3(x) = (1 - x)^2(1 + 2x)f_0 + x(1 - x)^2 f_0' + x^2(3 - 2x)f_1 + x^2(x - 1)f_1'.$$

Exercise 1.3 Use the result of Exercise 1.1 to obtain the coefficients in equation (1.5), and hence obtain the basis functions in equation (1.6).

Exercise 1.4 In the unit square, show that the basis functions at internal nodes are given by

$$\varphi_{i,j}(x, y) = \begin{cases} \left[\dfrac{x}{h} - (i - 1)\right]\left[\dfrac{y}{h} - (j - 1)\right] & \left(i - 1 \leqslant \dfrac{x}{h} \leqslant i; j - 1 \leqslant \dfrac{y}{h} \leqslant j\right) \\[2mm] \left[\dfrac{x}{h} - (i - 1)\right]\left[(j + 1) - \dfrac{y}{h}\right] & \left(i - 1 \leqslant \dfrac{x}{h} \leqslant i; j \leqslant \dfrac{y}{h} \leqslant j + 1\right) \\[2mm] \left[(i + 1) - \dfrac{x}{h}\right]\left[\dfrac{y}{h} - (j - 1)\right] & \left(i \leqslant \dfrac{x}{h} \leqslant i + 1; j - 1 \leqslant \dfrac{y}{h} \leqslant j\right) \\[2mm] \left[(i + 1) - \dfrac{x}{h}\right]\left[(j + 1) - \dfrac{y}{h}\right] & \left(i \leqslant \dfrac{x}{h} \leqslant i + 1; j \leqslant \dfrac{y}{h} \leqslant j + 1\right) \\[2mm] 0 & \text{elsewhere} \end{cases}$$

where $1 \leqslant i, j \leqslant m - 1$ and $mh = 1$.

Exercise 1.5 Consider the polynomial

$$g(x, y) = \sum_{r=0}^{3} \sum_{s=0}^{3} \alpha_{rs} x^r y^s$$

over the unit square $0 \leqslant x, y \leqslant 1$. Find the coefficients $\alpha_{rs}(0 \leqslant r, s \leqslant 3)$ in terms of the values of g, $\partial g/\partial x$, $\partial g/\partial y$, and $\partial^2 g/\partial x \partial y$ at the four corner points of the square. Show that the results of this calculation can be used to find the basis functions for the case $l = 2$ of the general theory of bivariate Hermite interpolation over a rectangular region subdivided into rectangular elements.

1.2 Function Spaces

This section contains an introduction to the mathematical structures required for an understanding of the theoretical developments of the finite element method. Only essential material will be presented and for a fuller treatment the interested reader is referred to Aubin (1977, 1979), Yosida (1971) or Milne (1980).

A *linear space* or *vector space* is a non-empty set X in which any two elements u and v can be combined, by a process called *addition*, to give some element in X denoted by $u + v$, provided the process of addition satisfies the following conditions:

(i) $u + v = v + u$
(ii) $u + (v + w) = (u + v) + w$
(iii) there exists a zero element φ such that $\varphi + u = u + \varphi = u$ for all u
(iv) for each u, there exists a negative $-u$ such that $u + (-u) = \varphi$.

It is also a necessary condition of a linear space that an element $u \in X$ can be combined with any real number or *scalar* α by *scalar multiplication* to give an element αu.

The process of scalar multiplication must satisfy the following conditions:

(v) $\alpha(u + v) = \alpha u + \alpha v$
(vi) $(\alpha + \beta)u = \alpha u + \beta u$
(vii) $(\alpha \beta)u = \alpha(\beta u)$
(viii) $1u = u$

One example of a linear space is the set of all N-dimensional real vectors, where $\mathbf{a} + \mathbf{b} = \mathbf{c}$ is defined by $a_i + b_i = c_i (i = 1, 2, \ldots, N)$ and $\alpha \mathbf{a} = \mathbf{d}$ by $d_j = \alpha a_j (j = 1, 2, \ldots, N)$.

A normed linear space (n.l.s.) is a linear space on which there is defined a norm $\|u\|$ such that

(i) $\|u\| \geqslant 0$
(ii) $\|u\| = 0 \Leftrightarrow u = 0$
(iii) $\|u + v\| \leqslant \|u\| + \|v\|$
(iv) $\|\alpha u\| = |\alpha| \|u\|$.

Thus we have the concept of the length of an element in the linear space. A *seminorm* satisfies (i), (iii), and (iv) but not (ii).

An inner product space (i.p.s.) or scalar product space is a linear space in which there is defined a real-valued function (u, v) for each pair of vectors in the linear space, such that

(i) $(\alpha u + \beta v, w) = \alpha(u, w) + \beta(v, w)$

(ii) $(u, v) = (v, u)$

(iii) $(u, u) > 0, \quad u \neq 0$.

Let $\{u_n\}$ be a sequence of points in an i.p.s.; then

(a) $\{u_n\}$ is a *Cauchy sequence* if for each $\varepsilon > 0$, there exists some $N = N(\varepsilon)$ such that for all $n, m \geq N$,

$$\| u_n - u_m \| < \varepsilon$$

(b) $\{u_n\}$ is a *convergent sequence* if there exists a point u in the i.p.s. such that for each $\varepsilon > 0$ there exists some $N = N(\varepsilon)$ such that for all $n \geq N$,

$$\| u_n - u \| < \varepsilon.$$

Exercise 1.6 Show that an i.p.s. is a n.l.s. with respect to the norm

$$\| u \| = (u, u)^{1/2}.$$

Then verify the parallelogram law

$$\| u + v \|^2 + \| u - v \|^2 = 2 \| u \|^2 + 2 \| v \|^2.$$

Show also that

$$4(u, v) = \| u + v \|^2 - \| u - v \|^2.$$

Exercise 1.7 Show that a convergent sequence is a Cauchy sequence.

1.2.1 Hilbert spaces

To show the converse of Exercise 1.7 is untrue consider the following. Let X be the space of points on the interval $(0, 1)$ and let $u_n = 1/n$. Then $\{u_n\}$ is a Cauchy sequence, but it is not a convergent sequence since the origin is not in the space.

A complete i.p.s. is one in which all Cauchy sequences are convergent sequences, and such a space is called a *Hilbert space*.

So far the spaces introduced have been such that a point in the space has represented a point on the real line, a vector, or a matrix. In order to provide a Hilbert space which is readily applicable to the development of finite element methods, it is necessary to introduce a space in which the points represent functions. The most useful function spaces can be developed from a simple Hilbert space denoted by $\mathcal{L}_2(R)$, where R is, for simplicity, an interval $[a, b]$ of the real line. Functions $f(x)$ are points in this space if and only if

$$\int_a^b f^2(x) \, dx$$

is finite. Such functions are said to be square integrable. For any two points $u(x)$

and $v(x)$, the inner product is defined by

$$(u, v) = \int_a^b u(x)v(x)\,dx \tag{1.16a}$$

and the norm by

$$\|u\|^2 = \int_a^b u^2(x)\,dx. \tag{1.16b}$$

Addition is defined by $(u + v)(x) = u(x) + v(x)$.

A subset of a linear space which is itself a linear space is called a *subspace*.

Exercise 1.8 Show that if $\|u\|$ and $\|v\|$ are finite, then (u, v) is also finite.

Exercise 1.9 Let $\mathcal{L}_2(R)$ be as previously defined. Which of the following subsets are subspaces?

(i) The functions u such that $u(a) = 1$.
(ii) The functions u such that $(du/dx)_{x=a} = 0$ and $u(b) = 0$.

Exercise 1.10 Let K be a proper subspace of the linear space $\mathcal{H}$, and let X be the set $\{u + h\}$, where u is some fixed point in $\mathcal{H}$ but not in K, and h is any point in K. Show that X, denoted by $\{u\} \oplus K$, is not a linear space.

The set X is called a *linear manifold*.

1.2.2 Linear operators

Let T be a mapping of the Hilbert space $\mathcal{H}_1$ into the Hilbert space $\mathcal{H}_2$. T is a linear mapping if and only if

(i) $T(u + v) = Tu + Tv$ (for all $u, v \in \mathcal{H}_1$)
(ii) $T(\alpha u) = \alpha Tu$ (for any scalar α).

A linear mapping T is said to be *bounded* if there exists a constant $M > 0$ such that

$$\|Tu\| \leqslant M\|u\| \quad \text{(for all } u \in \mathcal{H}_1\text{)}$$

and the smallest such M is called the norm of T and denoted by $\|T\|$; it follows that

$$\|T\| = \sup_{u \neq 0}\left\{\frac{\|Tu\|}{\|u\|}\right\} = \sup_{\substack{u \neq 0 \\ \|u\| \leqslant 1}}\left\{\frac{\|Tu\|}{\|u\|}\right\} = \sup_{\|u\| = 1}\{\|Tu\|\}. \tag{1.17}$$

A bounded linear mapping is *continuous*; for if the sequence $\{u_n\}$ has the limit point u then the sequence $\{Tu_n\}$ has the limit point T.

Let F be a bilinear mapping of $\mathcal{H}_1 \times \mathcal{H}_2$ into $\mathcal{H}_3$, that is, for any

$u_1 \in \mathscr{H}_1, u_2 \in \mathscr{H}_2, F(u_1, u_2) \in \mathscr{H}_3$ and F is linear in each of the arguments, then F is *bounded* if there exists $M > 0$ such that

$$\| F(u_1, u_2) \|_{\mathscr{H}_3} \leqslant M \| u_1 \|_{\mathscr{H}_1} \| u_2 \|_{\mathscr{H}_2}$$

and $\| F \|$ is the smallest such M. We shall denote by $\mathscr{L}(\mathscr{H}_1; \mathscr{H}_2)$ the space of bounded linear mappings from $\mathscr{H}_1$ into $\mathscr{H}_2$ and by $\mathscr{L}(\mathscr{H}_1 \times \mathscr{H}_2; \mathscr{H}_3)$ the space of bounded bilinear mappings from $\mathscr{H}_1 \times \mathscr{H}_2$ into $\mathscr{H}_3$. The space $\mathscr{L}(\mathscr{H}; \mathbb{R})$ of *bounded linear functionals* is called the *dual* of $\mathscr{H}$ and is denoted by $\mathscr{H}'$. Elements of $\mathscr{L}(\mathscr{H} \times \mathscr{H}; \mathbb{R})$ are referred to as *bilinear forms* (on $\mathscr{H}$).

A mapping T of $\mathscr{H}_1$ into $\mathscr{H}_2$ is denoted by

$$T: \mathscr{H}_1 \to \mathscr{H}_2$$

or

$$T: u \mapsto Tu$$

where the form Tu is preferred to $T(u)$ in general unless there is a danger of ambiguity. The alternative form is always used for both nonlinear mappings and linear functionals. A mapping $T: \mathscr{H} \to \mathscr{H}$ is called an *operator* on $\mathscr{H}$.

Theorem 1.1 Riesz representation theorem (Yosida, 1971; Showalter, 1977) $F \in \mathscr{H}'$ *if and only if there exists a unique* $v \in \mathscr{H}$ *such that for all* $u \in \mathscr{H}$,

$$F(u) = (u, v)$$

and

$$\| F \|_{\mathscr{H}'} = \| v \|_{\mathscr{H}}.$$

The mapping $J: F \mapsto v$ is called the *Riesz map*.

There is a 1–1 correspondence between $\mathscr{H}$ and $\mathscr{H}'$ defined by Theorem 1.1; such a correspondence is called an *isomorphism* and as the two spaces have the same structure they are said to be *isomorphic*. From the definition of an operator norm it follows that

$$\| F \|_{\mathscr{H}'} = \sup_{\| u \| \neq 0} \left\{ \frac{|F(u)|}{\| u \|_{\mathscr{H}}} \right\} \tag{1.18}$$

and from Theorem 1.1 it follows that there exists $v \in \mathscr{H}$ such that

$$F(u) = (u, v).$$

As it is possible to identify each element of $\mathscr{H}'$ with a unique element of $\mathscr{H}$ there is no ambiguity if we also write

$$\| v \|_{\mathscr{H}'} = \sup_{\| u \| \neq 0} \left\{ \frac{|(u, v)|}{\| u \|} \right\}.$$

A linear operator T which maps the whole of a Hilbert space $\mathscr{H}$ onto a particular subspace K is called a *projection* if and only if it maps the points of K onto themselves, i.e.

$$Tv = v \quad \text{(for all } v \in K).$$

If $A:\mathcal{H}_1 \rightarrow \mathcal{H}_2$ and $B:\mathcal{H}_2 \rightarrow \mathcal{H}_3$ the composite map $C:u \mapsto B(A(u))$ is denoted by

$$C = B \circ A.$$

A projection P is said to be an *orthogonal projection* if, for all u in the space, and all v in the subspace,

$$(u - Pu, v) = 0 \quad \text{(denoted by } (u - Pu) \perp v)$$

i.e. the remainder $u - Pu$ is orthogonal to all v in the subspace K. The remainder is said to belong to the orthogonal complement of K, which is denoted by $K^\perp$.

Lemma 1.1 If $\mathcal{H}$ is a Hilbert space and K is any subspace of $\mathcal{H}$, then $K^\perp$ is also a Hilbert space.

Lemma 1.2 If P is an orthogonal projection of a Hilbert space $\mathcal{H}$, onto some subspace K, then $I - P$ is an orthogonal projection onto $K^\perp$ and for any $u \in \mathcal{H}$, there exists $v \in K$ and $w \in K^\perp$ such that $u = v + w$ that is, $\mathcal{H} = K \oplus K^\perp$.

Theorem 1.2 The orthogonal projection P of a Hilbert space onto a subspace is unique, and the 'length' of the remainder $\| u - Pu \|$ is the minimum distance from u to the subspace.

Proof Assume that the orthogonal projection P is not unique. Then there exists another projection Q, such that

$$(u - Qu, v) = 0 \quad \text{(for all } v \in K).$$

Thus, since $Pu - Qu \in K$,

$$
\begin{aligned}
0 &= (u - Qu, Pu - Qu) \\
&= (u - Pu, Pu - Qu) + (Pu - Qu, Pu - Qu) \\
&= 0 + \| Pu - Qu \|^2 > 0
\end{aligned}
$$

since $Pu \neq Qu$, and so there is a contradiction. The original assumption is thus false, and so the orthogonal projection is unique.

Now let v be any point in K, other than Pu. Then as above

$$(v - Pu) \perp (u - Pu).$$

Hence

$$
\begin{aligned}
\| u - v \|^2 &= \| (u - Pu) + (Pu - v) \|^2 \\
&= \| u - Pu \|^2 + 2(u - Pu, Pu - v) + \| Pu - v \|^2 \\
&= \| u - Pu \|^2 + \| Pu - v \|^2 > \| Pu - u \|^2
\end{aligned}
$$

which is the desired result.

Corollary (i) $\| Pu \|^2 = (u, Pu)$ *and* $\| u - Pu \|^2 = \| u \|^2 - (u, Pu)$.

Corollary (ii) *For any* $u \in \mathcal{H}$ *there exists a unique* $u_0 \in K$ *and a unique* $v_0 \in K^\perp$ *such that* $u = u_0 + v_0$ *and*

$$\underset{v \in K}{\text{minimum}} \, \| u - v \|^2 = \| u - u_0 \|^2 = \| v_0 \|^2. \tag{1.19a}$$

It is possible to replace K by $K^\perp$ in (1.19a) to give a different minimum principle:

$$\underset{v \in K^\perp}{\text{minimum}} \, \| u - v \|^2 = \| u_0 \|^2.$$

Since

$$\| u \|^2 = \| u_0 \|^2 + \| v_0 \|^2$$

we can replace this minimum principle by a *maximum* principle

$$\underset{v \in K^\perp}{\text{minimum}} \, \{ \| u \|^2 - \| u - v \|^2 \}$$

or

$$\underset{v \in K^\perp}{\text{maximum}} \, \{ 2(u, v) - \| v \|^2 \} \tag{1.19b}$$

which has the same solution as (1.19a).

A linear operator T is said to be *positive definite* if and only if there exists $\gamma > 0$ such that

$$(Tu, u) \geqslant \gamma \| u \|^2 \quad \text{(for all } u \neq 0)$$

or *positive semi-definite* if

$$(Tu, u) \geqslant 0 \quad \text{(for all } u \neq 0).$$

For example, the mapping $\begin{pmatrix} \alpha \\ \beta \end{pmatrix}$ to $\begin{pmatrix} \alpha/2 \\ \beta/2 \end{pmatrix}$ is positive definite ($\gamma = \tfrac{1}{2}$), whereas the mapping $\begin{pmatrix} \alpha \\ \beta \end{pmatrix}$ to $\begin{pmatrix} \alpha \\ 0 \end{pmatrix}$ is positive semi-definite, since

$$(Tu, u) = \alpha^2$$

which attains the value zero whenever $\alpha = 0$, for arbitrary β.

The *adjoint* of T is the operator T^* such that

$$(Tu, v) = (u, T^*v) \quad \text{(for all } u, v).$$

If $(Tu, v) = (u, Tv)$, then T is *self-adjoint*.

A bilinear form F on $\mathcal{H}$ is said to be *positive definite* (or *coercive* or $\mathcal{H}$-*elliptic*) if and only if there exists $\gamma > 0$ such that

$$F(u, u) \geqslant \gamma \| u \|^2.$$

If the bilinear form F is *symmetric*, i.e.

$$F(u, v) = F(v, u) \quad \text{for all } u, v \in \mathscr{H}$$

as well as positive definite, then it can be used to define an alternative inner product on the space $\mathscr{H}$. We say that the components of the space $\mathscr{H}$, equipped with this new *energy inner product*

$$(u, v)_F = F(u, v)$$

form the *energy space* $\mathscr{H}_F$ and the corresponding *energy norm* is

$$\| u \|_F = [F(u, v)]^{1/2}.$$

A linear operator T such that $\| T \| < 1$ is known as a *contraction mapping*.

Theorem 1.3 *If T is a bounded positive definite linear operator then the bounded linear operator T^{-1} exists.*

Proof Assume w.l.o.g. that $\| T \| < 1$ since it is always possible to scale the operator to satisfy this inequality and if $(\lambda T)^{-1}$ exists, clearly it follows that T^{-1} exists.

As T is positive definite, there exists γ such that

$$(Tu, u) \geqslant \gamma \| u \|^2 \quad (1 > \gamma > 0).$$

Then

$$((I - T)u, u) \leqslant (1 - \gamma) \| u \|^2$$

thus

$$\| I - T \| \leqslant (1 - \gamma) < 1.$$

It follows that the sequence

$$S_n = \sum_{i=0}^{n} (I - T)^i$$

converges in the norm. Thus there exists

$$S = \lim_{n \to \infty} \sum_{i=0}^{n} (I - T)^i$$

but

$$(I - (I - T))S = I - \lim_{n \to \infty} (I - T)^{n+1} = I.$$

Thus $S = (I - (I - T))^{-1} = T^{-1}$ which must therefore exist.

Corollary   It follows from Theorem 1.3 that there is a unique solution $u \in \mathscr{H}$ of the equation

$$Tu = w \qquad (1.20a)$$

for any given $w \in \mathscr{H}$ provided that T is a bounded positive definite linear operator.

Frequently an operator equation is posed in the alternative *weak formulation*: find $u \in \mathscr{H}$ such that

$$a(u, v) = F(v) \quad \text{for all } v \in \mathscr{H} \qquad (1.20b)$$

where a is a bounded, positive definite bilinear form on $\mathscr{H}$ and F is a linear functional on $\mathscr{H}$, i.e. $F \in \mathscr{H}'$. The corollary above, together with Exercise 1.12, proves the existence of a solution to (1.20b), but existence and uniqueness are usually stated together as a standard result.

Theorem 1.4 Lax–Milgram lemma (Ciarlet, 1978; Showalter, 1977). If a is a bounded, positive definite bilinear form on a Hilbert space $\mathscr{H}$ and $F \in \mathscr{H}'$ then there exists a unique $u \in \mathscr{H}$ such that (1.20b) is satisfied.

Proof   For each $u \in \mathscr{H}$ it follows that $a(u, \ ) \in \mathscr{H}'$. We denote this functional by G, i.e.

$$G(v) \equiv a(u, v) \quad \text{for all } v \in \mathscr{H}.$$

From the Reisz representation theorem, there exists a $g \in \mathscr{H}$ such that

$$(g, v) = G(v) \quad \text{for all } v \in \mathscr{H}.$$

The mapping $A : u \mapsto G$ is linear and the Reisz map $J : G \mapsto g$ is linear, thus the composite mapping $T : u \mapsto Tu \equiv g$ is also linear and it follows that

$$a(u, v) = (Tu, v) \quad \text{for all } v \in \mathscr{H}.$$

Thus (1.20b) can be rewritten as

$$(Tu, v) = (w, v) \quad \text{for all } v \in \mathscr{H} \qquad (1.20c)$$

where, using the Reisz map a second time, $w = J(F)$. As $\mathscr{H}$ is complete, (1.20c) implies that

$$Tu = w$$

and thus we have derived the equivalence of the two forms (1.20a) and (1.20b). The uniqueness of the solution follows from Theorem 1.3.

In Chapter 4, mention is made of nonlinear problems. A nonlinear operator T is said to be *monotone* if

$$(T(u) - T(v), u - v) \geqslant 0 \quad \forall u, v \in \mathscr{H}$$

and *strictly monotone* if

$$(T(u) - T(v), u - v) \geqslant \gamma \|u - v\|^2 \quad \forall u, v \in \mathcal{H}$$

It is possible to prove a result similar to Theorem 1.3 for the existence and uniqueness of solutions to

$$T(u) = v$$

for strictly monotone operators T (see Varga, 1971; Showalter, 1977).

Exercise 1.11 Let $E \in \mathcal{L}(\mathcal{H}_1 \times \mathcal{H}_2; \mathcal{H}_3)$ and define $F \in \mathcal{L}(\mathcal{H}_1; \mathcal{H}_3)$ such that for some fixed $v \in \mathcal{H}_2$

$$Fu = E(u, v) \quad \text{(for all } u \in \mathcal{H}_1).$$

Prove that

$$\|F\| \leqslant \|E\| \|v\|_{\mathcal{H}_2}.$$

Note that throughout this book subscripts may be omitted from operator norms in the interest of brevity.

Exercise 1.12 Prove, using the Riesz representation theorem, that given any continuous bilinear form a, there exist unique operators $M, N \in \mathcal{L}(\mathcal{H}; \mathcal{H})$ that satisfy

$$a(u, v) = (Mu, v) = (u, Nv).$$

Note that $N = M^*$.

Exercise 1.13 Prove Lemma 1.1 and Lemma 1.2.

Exercise 1.14 Which of the following linear operators are projections?

(i) T which maps the 2-vector $\begin{pmatrix} \alpha \\ \beta \end{pmatrix}$ into the vector $\begin{pmatrix} \alpha \\ 0 \end{pmatrix}$.

(ii) Q which maps the space $\mathcal{L}_2(R), R = [a, b]$ onto the space of linear functions, by linear interpolation at the endpoints.

(iii) S which maps the space of 2×2 matrices into diagonal matrices by

$$S\left(\begin{bmatrix} \alpha & \beta \\ \gamma & \delta \end{bmatrix} \right) = \begin{bmatrix} \alpha + \delta & \\ & \beta + \gamma \end{bmatrix}.$$

(iv) S^2, with S as in (iii).

Exercise 1.15 Prove that if

$$(Tu, u) \leqslant \gamma \|u\|^2$$

then

$$\|T\| \leqslant \gamma.$$

1.2.3 Approximating subspaces

Let $\mathcal{H}$ be a space and let $S = \{\varphi_1, \ldots, \varphi_N\}$ be a set of N elements of $\mathcal{H}$. It is assumed that S is a linearly independent set, i.e. there is no set of scalars

$\alpha_i (i = 1, 2, \ldots, N)$, excluding the zero set $\alpha_1 = \alpha_2 = \cdots = \alpha_N = 0$, such that

$$\alpha_1 \varphi_1 + \alpha_2 \varphi_2 + \cdots + \alpha_N \varphi_N = 0.$$

Then if for every element $u \in \mathscr{H}$ there exist scalars $\beta_i (i = 1, 2, \ldots, N)$ such that

$$u = \beta_1 \varphi_1 + \beta_2 \varphi_2 + \cdots + \beta_N \varphi_N$$

the set S is said to form a *basis* for $\mathscr{H}$, which is said to be an *N-dimensional space.*

For example, the space of vectors of the form $\begin{pmatrix} \alpha \\ \beta \end{pmatrix}$ is a two-dimensional

space. $S_1 = \left\{ e_1 = \begin{pmatrix} 0 \\ 1 \end{pmatrix}, \ e_2 = \begin{pmatrix} 1 \\ 0 \end{pmatrix} \right\}$ forms a basis as does $S_2 = \left\{ f_1 = \begin{pmatrix} 1 \\ 1 \end{pmatrix}, \right.$

$\left. f_2 = \begin{pmatrix} 1 \\ -1 \end{pmatrix} \right\}$. These two bases are illustrated in Figure 1.10.

The importance of finite-dimensional function spaces becomes apparent when the approximating functions mentioned in Section 1.1 are considered as elements of $\mathscr{L}_2(R)$. For example, if the interval $R = [a, b]$ is partitioned by the points $x_i (i = 0, 1, \ldots, n)$ it is possible to construct the set of Hermite functions which are piecewise linear on the interval. Any linearly independent set of $(n + 1)$ such functions constitutes a basis for $H^{(1)}(\Pi, R)$. It is a simple matter to show that H is a complete subspace of $\mathscr{L}_2(R)$ (the details are left as an exercise for the reader) and so H is an $(n + 1)$-dimensional subspace of $\mathscr{L}_2(R)$. In fact the pyramid functions given by (1.3) constitute a natural choice of basis for H in view of the significance of the parameters

$$f_i \quad (i = 0, 1, \ldots, n).$$

Once an approximation to any $f \in \mathscr{L}_2(R)$ has been constructed, it is reasonable

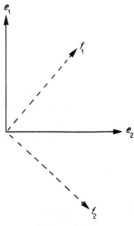

Figure 1.10

to ask the question 'How good is the approximation?'. The approximation is *best* if the element $\tilde{f} \in K$ is such that the length of the error $f - \tilde{f}$ is minimized. In Theorem 1.2 it was shown that if P is the orthogonal projection onto K, then $\|f - Pf\|$ is the minimum distance from f to the subspace K, and hence the best approximation is $\tilde{f} = Pf$. Since it is assumed that K is of finite dimension, this can be used to calculate the best approximate. Assume that the set $S = \{f_1, \ldots, f_N\}$ forms a basis for K. Then since

$$(f - \tilde{f}, g) = 0$$

for all $g \in K$, it follows that

$$\left(f - \tilde{f}, \sum_{k=1}^{N} \beta_k f_k \right) = 0$$

for all possible sets of scalars $\beta_k (k = 1, \ldots, N)$, and so

$$(f - \tilde{f}, f_k) = 0 \quad (k = 1, \ldots, N). \tag{1.21}$$

If $\tilde{f} = \sum_{i=1}^{N} \alpha_i f_i$, it follows that

$$(f, f_k) - \sum_{i=1}^{N} \alpha_i (f_i, f_k) = 0 \quad (k = 1, \ldots, N).$$

This can be written as

$$G\alpha = \mathbf{b} \tag{1.22}$$

where

$$G = [g_{ik}] = [(f_i, f_k)]$$
$$\alpha = \{\alpha_1, \ldots, \alpha_N\}^{\mathrm{T}}$$

and

$$\mathbf{b} = \{(f, f_1), \ldots, (f, f_N)\}^{\mathrm{T}}.$$

The matrix G is called the Gram matrix, and the equations (1.22) are called *normal equations*.

Methods of approximation involving the solution of a set of normal equations to obtain the orthogonal projection of a function onto a finite-dimensional subspace are referred to as *projection* methods of approximation.

Exercise 1.16 Prove that any two bases of the some space must have the same number of elements.

Exercise 1.17 Give a basis for the space of cubic spline functions and the space of piecewise planar functions defined on a triangular mesh.

The approximating functions mentioned in Section 1.1 are all particular cases of a general approximation problem. This problem is to associate with every element f, of a Hilbert space $\mathscr{H}$, a unique element $\tilde{f}$, of an N-dimensional approximating subspace K (for example, if $K = H$ then $N = n + 1$). Then $\tilde{f}$ is said to be the K-approximate of f. The mapping of f into $\tilde{f}$ is usually a linear mapping and a projection. For example if $K = H$, $\tilde{f}$ can be found by linear interpolation between the partition points. This is a unique mapping which is also a projection.

Exercise 1.18 Verify that for $\mathscr{H} = \mathscr{L}_2(R)$, with the inner product and norm given by (1.17) and (1.18) respectively, the best approximate $\tilde{f}$ as defined by (1.21) is also best in the least squares sense. Note that the normal equations (1.22) are not recommended for calculating the solution to a least squares problem since they tend to be ill conditioned. (1.17) and (1.18) respectively, the best approximate $\tilde{f}$ as defined by (1.21) is also best in the

Exercise 1.19 Verify that it is possible to define the Hilbert space $\mathscr{L}_2^w(R)$ using the inner product

$$(u, v)_w = \int_a^b w(x)u(x)v(x)\,\mathrm{d}x$$

where $w(x) \geqslant 0$ is a suitable weight function. State the normal equations which determine the best weighted least squares approximation of a function $f(x)$ in $\mathscr{L}_2^w(R)$.

Exercise 1.20 For functions $u(x)$ and $v(x)$ that are square integrable and have a square integrable first derivative, verify that it is possible to define the Hilbert space $\mathscr{H}_2^{(1)}(R)$ using the inner product

$$(u, v)_1 = \int_a^b \{u(x)v(x) + u'(x)v'(x)\}\,\mathrm{d}x$$

and the norm

$$\|u\|_1^2 = \int_a^b \{u^2(x) + u'^2(x)\}\,\mathrm{d}x$$

where the prime indicates differentiation with respect to x. Derive the normal equations for best approximation of a function $f(x)$ in this space. The space $\mathscr{H}_2^{(1)}(R)$ is an example of a *Sobolev* space.

1.2.4 Sobolev spaces

It is useful at this stage to introduce a more general definition of Sobolev spaces of functions defined in any number of dimensions. It will be shown in later chapters that the most natural error bounds are defined in terms of Sobolev norms.

A *multi-index* i is a vector of non-negative integers $i = (i_1, \ldots, i_m)$ and a partial derivative in $\mathbb{R}^m$ can be denoted by

$$D^i u = \frac{\partial^{|i|}}{\partial x_1^{i_1} \cdots \partial x_m^{i_m}} u \tag{1.23}$$

where

$$|i| = i_1 + \cdots + i_m.$$

Thus

$$\sum_{|i|=k} D^i u$$

includes all derivatives of order k and

$$\sum_{|i| \leqslant k} D^i u$$

includes the function value and all derivatives up to and including order k.

A Sobolev semi-norm is defined as

$$|u|^2_{k,R} = \int \cdots \int_R \sum_{|i|=k} (D^i u)^2 \, d\mathbf{x}$$

while a *Sobolev norm* is

$$\|u\|^2_{k,R} = \sum_{l=0}^{k} |u|^2_{l,R}.$$

The *Sobolev space* $\mathscr{H}^{(k)}(R)$ is

$$\mathscr{H}^{(k)}(R) = \{u : \|u\|^2_{k,R} < \infty\}. \tag{1.24}$$

The Sobolev space $\mathscr{\mathring{H}}^{(k)}_2(R)$ is

$$\mathscr{\mathring{H}}^{(k)}(R) = \{u \in \mathscr{H}^{(k)}(R) : D^\alpha u = 0 \text{ on } \partial R, |\alpha| < k\}.$$

An alternative definition of $\mathscr{\mathring{H}}^{(k)}(R)$ is the closure of the space $\mathscr{D}(R)$ in the norm $\|u\|_{k,R}$. The space $\mathscr{D}(R)$ is defined as the space of continuously differentiable functions that vanish in a neighbourhood of ∂R, i.e. that have compact support in R.

The index k clearly indicates the degree of smoothness necessary for a function to be an element of $\mathscr{H}^{(k)}$. It is obvious from the definition that

$$\mathscr{H}^{(k)}(R) \supset \mathscr{H}^{(k+1)}(R)$$

and that

$$\|u\|_{k,R} \leqslant \|u\|_{k+1,R}.$$

In general, functions $u \in \mathscr{\mathring{H}}^{(k)}_2(R)$ can possess slightly less smoothness on the boundary ∂R (a set of measure zero) than is required within the interior of R.

It is possible to quantify this requirement in terms of a Sobolev space $\mathscr{H}^{(k-1/2)}_2(\partial R)$ (i.e. a fractional index). The norm is then defined as

$$\|v\|_{k-1/2,\partial R} = \inf_{u \in \mathscr{H}^{(k)}_2(R)} \{u : u = v \text{ on } \partial R\}. \tag{1.25}$$

The reason for the choice $k - \frac{1}{2}$ rather than any other value between k and $k - 1$ is beyond the scope of this book. Those interested should consult Nečas (1967), Lions and Magenes (1972), or Aubin (1979).

The dual of the space $\mathscr{H}_2^{(k)}(R)$ is denoted by $\mathscr{H}_2^{(-k)}(R)$.

1.2.5 Green's theorem (or Green's formula)

The use of integration by parts, i.e. Green's theorem, to transform the bilinear form $a(u, v)$ into an $\mathscr{L}_2$ inner product (Tu, v) as required in the Lax–Milgram lemma, is fundamental to the success of the Galerkin method.

Theorem 1.5 (Green's theorem) *Possibly the simplest statement of Green's theorem for the purpose of this book is that if $u, v \in \mathscr{H}^{(1)}(R)$ then*

$$\int \ldots \int_R \left(u \frac{\partial v}{\partial x_i} + v \frac{\partial u}{\partial x_i} \right) dx = \int \ldots \int_{\partial R} u v n_i \, ds \qquad (1.26)$$

where n_i is the component of the unit outward normal in the coordinate direction corresponding to the variable x_i.

Proof We shall provide a proof for the case of simple two-dimensional regions only; it should be obvious to the reader how to generalize the method. We assume that the boundary ∂R, on the domain R, appears as in Figure 1.11, with a single highest point $y = b$ and a single lowest point $y = a$. The boundary can then be split into a left-hand-side arc ∂R_l in the form $x = X_l(y)$ and a right-hand-side arc ∂R_r of the form $x = X_r(y)$. Local maxima add unnecessary complications to

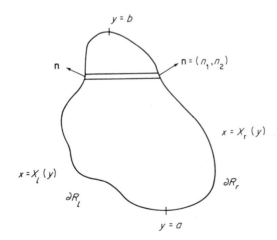

Figure 1.11

the notation used in the proof. Then any function $w(x, y) \in \mathcal{H}^{(1)}(R)$ satisfies

$$\iint_R \frac{\partial w}{\partial x} \, dx \, dy = \int_a^b \int_{X_l(y)}^{X_r(y)} \frac{\partial w}{\partial x} \, dx \, dy$$

$$= \int_a^b \{w(X_r(y), y) - w(X_l(y), y)\} \, dy. \qquad (1.2.7)$$

On the arc $x = X_r(y)$, $dy = n_1 \, ds$ where s is arc length, while on $x = X_l(y)$, $dy = -n_1 \, ds$. Thus (1.27) can be written as

$$\iint_R \frac{\partial w}{\partial x} \, dx \, dy = \int_{\partial R_r} w n_1 \, ds + \int_{\partial R_l} w n_1 \, ds = \int_{\partial R} w n_1 \, ds.$$

Thus if $w = uv$ we have that

$$\iint_R \left\{ u \frac{\partial v}{\partial x} + v \frac{\partial u}{\partial x} \right\} dx \, dy = \int_{\partial R} uv n_1 \, ds.$$

The proof can be repeated using $\partial w / \partial y$ and hence we have the desired result, at least in $\mathbb{R}^2$.

2

Variational Principles

2.1 Introduction

Variational principles occur widely in physical and other problems and approximate methods of solution of such problems are often based on associated variational principles. The mathematical formulation of a variational principle is that the integral of some typical function has a smaller (or larger) value for the actual performance of the system than for any virtual performance subject to the general conditions of the system. The integrand is a function of coordinates, field amplitudes, and their derivatives and the integration is over a region governed by the coordinates of the system, which may include the time. The problem of determining the minimum of the integral often leads to one or more partial differential equations together with appropriate boundary conditions. It is not the intention of this book to give approximate methods of solution of these differential equations as a means of solving the original physical problems formulated as variational principles. Instead we intend to outline an approximate method which is directly based on the variational principle.

As an example of determining the extremum of an integral, consider the double integral

$$I(u) = \iint_R F(x, y, u, u_x, u_y) \, dx \, dy \tag{2.1}$$

where u is continuous and has continuous derivatives up to the second order, and takes prescribed values on the boundary of R, which is a bounded region in the (x, y) plane. It is relatively easy (Courant and Hilbert, 1953) to see that the necessary condition for $I(u)$ to have an extremum is that $u(x, y)$ must satisfy the *Euler–Lagrange* differential equation

$$\frac{\partial}{\partial x} F_{u_x} + \frac{\partial}{\partial y} F_{u_y} - F_u = 0. \tag{2.2}$$

27

From the many solutions of this equation, the particular solution is selected which satisfies the given boundary conditions. For example, when

$$F = \tfrac{1}{2}(u_x^2 + u_y^2)$$

(2.2) reduces to Laplace's equation,

$$\frac{\partial^2 u}{\partial x^2} + \frac{\partial^2 u}{\partial y^2} = 0.$$

The reason for demanding continuous second derivatives for u is now clear; it is to justify the existence of the Euler equation. Approximate methods based on the minimization of I in (2.1), however, only require continuity of u and piecewise continuity of the first derivatives.

We now return to the basic problem of the variational principle which is to determine the function from an admissible class of functions such that a certain definite integral involving the function and some of its derivatives takes on a maximum or minimum value in a closed region R. This is a generalization of the elementary theory of maxima and minima of the calculus which is concerned with the problem of finding a point in a closed region at which a function has a maximum or minimum value compared with neighbouring points in the region. The definite integral in the variational principle is an example of a *functional* and it depends on the entire course of a function rather than on a number of variables. The domain of the functional is the space of the admissible functions. The main difficulty with the variational principle approach is that problems which can be meaningfully formulated as variational principles may not have solutions. This is reflected in mathematical terms by the domain of admissible functions of the functional not forming a closed set. *Thus the existence of an extremum (maximum or minimum) cannot be assumed for a variational principle.* However, in this text we are concerned with *approximate* solutions of variational principles. These are obtained by considering some closed subset of the space of admissible functions to provide an upper and lower bound for the theoretical solution of the variational principle.

One apparent advantage of the variational approach is that we seem to require less continuity in the solution function. This paradox is explained at length on pages 199–204 of Courant and Hilbert (1953) and Chapter 2 of Clegg (1967). As a consequence of the weaker continuity requirements, a useful advantage of the variational approach is the greater ease with which approximate solutions can be constructed. A large part of the present text is devoted to describing such approximate methods.

The space over which the integral is evaluated in a variational principle may contain the time coordinate. We shall look first at variational principles which do not involve the time. These variational principles, usually of minimum potential energy type, govern problems of stable equilibrium which arise from classical field problems of mathematical physics. This chapter contains only the material on variational principles which is relevant to the main theme of this book. No proofs or detailed discussions are given and the interested reader is referred to appropriate parts of books such as Courant and Hilbert (1953), Morse and Feshbach (1953), Hildebrand (1965), Schechter (1967), Clegg (1967), and Oden and Reddy (1976).

2.2 Stable Equilibrium Problems

The differential equation which is associated with a variational principle is known as the *Euller–Lagrange* equation. It is a necessary, but rarely sufficient, condition which a function must satisfy if it is to maximize or minimize a definite integral. The simplest problem of the variational calculus is to determine the minimum of the integral

$$I(u) = \int_{x_0}^{x_1} F(x, u(x), u'(x)) \, dx$$

where the values $u(x_0)$ and $u(x_1)$ are given, and a prime denotes differentiation with respect to x. The necessary but not sufficient condition for the minimum to exist is that $u(x)$ satisfies the differential equation

$$\frac{\partial F}{\partial u} - \frac{d}{dx} \frac{\partial F}{\partial u'} = 0.$$

We shall now generalize this result to cover the following cases:

(1) *Two dependent variables* The integral to be minimized is

$$I(u, v) = \int_{x_0}^{x_1} F(x, u(x), v(x), u'(x), v'(x)) \, dx$$

where the values $u(x_0), u(x_1), v(x_0), v(x_1)$ are given. The necessary conditions are

$$\frac{\partial F}{\partial u} - \frac{d}{dx} \frac{\partial F}{\partial u'} = 0$$

$$\frac{\partial F}{\partial v} - \frac{d}{dx} \frac{\partial F}{\partial v'} = 0.$$

(2) *Two independent variables* The integral is

$$I(u) = \iint_{R} F(x, y, u(x, y), u_x(x, y), u_y(x, y)) \, dx \, dy$$

where u takes on prescribed values on the boundary of R, the region of integration. The necessary condition is

$$\frac{\partial F}{\partial u} - \frac{\partial}{\partial x}\frac{\partial F}{\partial u_x} - \frac{\partial}{\partial y}\frac{\partial F}{\partial u_y} = 0.$$

(3) *Higher derivatives*   For variational principles involving second derivatives, the integral is

$$I(u) = \int_{x_0}^{x_1} F(x, u(x), u'(x), u''(x))\, dx$$

where the values $u(x_0)$, $u'(x_0)$, $u(x_1)$, $u'(x_1)$ are given and the corresponding necessary condition is

$$\frac{\partial F}{\partial u} - \frac{d}{dx}\frac{\partial F}{\partial u'} + \frac{d^2}{dx^2}\frac{\partial F}{\partial u''} = 0.$$

(4) *Constrained extrema*   Here the variational problem is constrained by one or more auxiliary conditions. One integral expression is to be made an extremum (maximum or minimum) while one or more other integral expressions maintain fixed values. Such problems are termed *isoperimetric* problems. Consider for example the problem of determining the function $u(x)$ which maximizes (or minimizes) the integral

$$I(u) = \int_{x_0}^{x_1} F(x, u(x), u'(x))\, dx$$

subject to the condition that $u(x)$ satisfies the equation

$$\int_{x_0}^{x_1} G(x, u(x), u'(x))\, dx = \alpha \tag{2.3}$$

where the constant α is given. The necessary condition for an extremum to exist is that

$$\frac{\partial(F + \lambda G)}{\partial u} - \frac{d}{dx}\frac{\partial(F + \lambda G)}{\partial u'} = 0$$

where the numerical value of the parameter λ is chosen so that (2.3) is satisfied. A simple example of an isoperimetric problem is the catenary. The problem is to find the shape of a uniform heavy string with fixed end points which hangs under gravity. Here we require to find the function $u(x)$ which passes through the points (x_0, u_0) and (x_1, u_1) and makes the integral

$$\int_{x_0}^{x_1} u(1 + u'^2)^{1/2}\, dx$$

as small as possible, while maintaining a fixed value for the integral

$$\int_{x_0}^{x_1} (1 + u'^2)^{1/2}\, dx.$$

Before proceeding further, we give some examples of variational principles and equivalent Euler–Lagrange equations. In these examples the region under consideration is R with ∂R as its boundary.

(1) *Dirichlet problem for Laplace's equation*

$$I(u) = \iint_R \tfrac{1}{2}(u_x^2 + u_y^2)\, dxdy \quad (u \text{ given on } \partial R)$$

$$u_{xx} + u_{yy} = 0.$$

(2) *Loaded and clamped plate* (Biharmonic operator)

$$I(u) = \iint_R \tfrac{1}{2}[u_{xx}^2 + 2u_{xy}^2 + u_{yy}^2 - 2qu]\, dxdy \quad \left(u = \frac{\partial u}{\partial n} = 0 \text{ on } \partial R\right)$$

$$u_{xxxx} + 2u_{xxyy} + u_{yyyy} = q(x, y)$$

$q(x, y)$ is the normal load on the plate.

(3) *Small displacement theory of elasticity* (Plane stress, applicable to very thin plates)

$$I(u, v) = \iint_R \tfrac{1}{2}[(1 - v)(u_x^2 + v_y^2) + v(u_x + v_y)^2$$

$$+ \tfrac{1}{2}(1 - v)(u_y + v_x)^2]\, dxdy \quad (u, v \text{ given on } \partial R)$$

$$u_{xx} + vv_{xy} + \tfrac{1}{2}(1 - v)(u_{yy} + v_{xy}) = 0$$

$$v_{yy} + vu_{xy} + \tfrac{1}{2}(1 - v)(u_{xy} + v_{xx}) = 0.$$

(4) *Radiation (e^u) and molecular diffusion (u^2)*

$$I(u) = \iint_R \tfrac{1}{2}\left(u_x^2 + u_y^2 + \begin{cases} 2e^u \\ \tfrac{2}{3}u^3 \end{cases}\right) dxdy \quad (u \text{ given on } \partial R)$$

$$u_{xx} + u_{yy} = \begin{cases} e^u \\ u^2 \end{cases}.$$

(5) *Plateau's problem* (To find the surface of minimum area bounded by a closed curve in three-dimensional space)

$$I(u) = \iint_R \tfrac{1}{2}(1 + u_x^2 + u_y^2)^{1/2}\, dxdy \quad (u \text{ given on } \partial R)$$

$$\nabla(\gamma_1 \nabla u) = 0 \quad \gamma_1 = (1 + u_x^2 + u_y^2)^{-1/2}.$$

(6) *Non-Newtonian fluids*

$$I(u) = \iint_R [\tfrac{1}{2}(u_x^2 + u_y^2)^{1+\varepsilon} + uc]\,dxdy \quad (u \text{ given on } \partial R)$$

$$\nabla(\gamma_2 \nabla u) = c \,(\text{constant}) \quad \gamma_2 = (u_x^2 + u_y^2)^\varepsilon \quad (0 \geqslant \varepsilon \geqslant -\tfrac{1}{2}).$$

(7) *Compressible flow*

$$I(p) = \iint_R p\,dx$$
$$(\rho\nabla\phi) = 0$$

$\begin{cases} \rho \text{ density,} \\ p \text{ pressure,} \\ \phi \text{ velocity potential.} \end{cases}$

Of the problems above, the first three are linear, the fourth is mildly nonlinear, and the last three are nonlinear.

Exercise 2.1 Show that the length of the curve connecting two points (x_0, u_0) and (x_1, u_1) is

$$I(u) = \int_{x_0}^{x_1} (1 + u'^2)^{1/2}\,dx.$$

Use the associated Euler–Lagrange equation to find the shortest path between the two points.

Exercise 2.2 Find the function $u(x)$ which passes through the points (x_0, u_0) and (x_1, u_1) and gives the minimum surface of revolution when rotated about the x-axis.

Exercise 2.3 Show that

$$\frac{\partial^2 u}{\partial x^2} + \frac{\partial^2 u}{\partial y^2} + \frac{\partial^2 u}{\partial z^2} + f(x, y, z) = 0$$

is the necessary condition for the integral

$$I(u) = \iiint_R \tfrac{1}{2}[u_x^2 + u_y^2 + u_z^2 - 2uf(x, y, z)]\,dxdydz$$

to be a minimum when u is specified on the surface ∂R which surrounds the volume R.

2.3 Boundary Conditions

The steady problems discussed in the previous section have been such that the function is specified on the boundary and so is not subject to variation there. In many problems, however, the function is not specified on the boundary and alternative equally valid boundary conditions apply. Consider, for example (Courant and Hilbert, 1953, pp. 208, 209), the variational problem consisting of the minimization of the integral

$$I(u) = \int_{x_0}^{x_1} F(x, u, u') \, dx \tag{2.4}$$

where u is not specified at the boundary points $x = x_0, x_1$. The necessary conditions for u to minimize $I(u)$ are that u satisfies the Euler–Lagrange equation

$$\frac{\partial F}{\partial u} - \frac{d}{dx} \frac{\partial F}{\partial u'} = 0$$

together with the boundary conditions

$$\frac{\partial F}{\partial u'} = 0 \quad (x = x_0, x_1).$$

The latter are known as *natural* boundary conditions because they follow directly from the minimization of the basic integral. If the boundary conditions of the problem consist of neither u specified at the boundary points nor the natural boundary conditions, then the functional to be minimized must be modified in an appropriate manner.

Consider the following modified form of (2.4):

$$I(u) = \int_{x_0}^{x_1} F(x, u, u') \, dx + [g_1(x, u)]_{x = x_1} - [g_0(x, u)]_{x = x_0}$$

where $g_0(x, u)$ and $g_1(x, u)$ are unspecified functions. The necessary conditions for this modified integral to have a minimum are (Schechter, 1967, p. 28)

$$\frac{\partial F}{\partial u} - \frac{d}{dx} \frac{\partial F}{\partial u'} = 0$$

together with the boundary conditions

$$\frac{\partial F}{\partial u'} + \frac{\partial g_0}{\partial u} = 0 \quad (x = x_0)$$

and

$$\frac{\partial F}{\partial u'} + \frac{\partial g_1}{\partial u} = 0 \quad (x = x_1).$$

Thus the functions $g_0(x, u)$ and $g_1(x, u)$ can be obtained to suit the boundary conditions of the problem. For example, the variational principle equivalent to the differential problem consisting of the equation

$$u'' + f(x) = 0$$

together with the boundary conditions

$$-u' + \alpha u = 0 \quad (x = x_0)$$

and

$$u' + \beta u = 0 \quad (x = x_1)$$

is based on the functional

$$I(u) = \int_{x_0}^{x_1} [\tfrac{1}{2}u'^2 - f(x)u]\,dx + [\tfrac{1}{2}\beta u^2]_{x=x_1} - [\tfrac{1}{2}\alpha u^2]_{x=x_0}.$$

If we now consider a variational problem in two space dimensions consisting of the minimization of the integral

$$I(u) = \iint_R F(x, y, u, u_x, u_y)\,dx\,dy$$

where u is not specified on the boundary of the region R, the necessary conditions for u to minimize $I(u)$ are that u satisfies that differential equation

$$\frac{\partial F}{\partial u} - \frac{\partial}{\partial x}\frac{\partial F}{\partial u_x} - \frac{\partial}{\partial y}\frac{\partial F}{\partial u_y} = 0$$

together with the natural boundary condition

$$\frac{\partial F}{\partial u_x}\frac{dy}{ds} - \frac{\partial F}{\partial u_y}\frac{dx}{ds} = 0$$

on the curve ∂R which encloses the region R. If the normal to ∂R makes an angle α with the x-axis (see Figure 2.1), then $\cos\alpha = dy/ds$, and $\sin\alpha = -dx/ds$, where s denotes arc length along the boundary. If this is extended to the case of two dependent functions u and v (Hildebrand, 1965, p. 135), the additional Euler–Lagrange equation in v is

$$\frac{\partial F}{\partial v} - \frac{\partial}{\partial x}\frac{\partial F}{\partial v_x} - \frac{\partial}{\partial y}\frac{\partial F}{\partial v_y} = 0$$

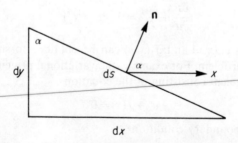

Figure 2.1

and the additional natural boundary condition is

$$\frac{\partial F}{\partial v_x}\frac{dy}{ds} - \frac{\partial F}{\partial v_y}\frac{dx}{ds} = 0.$$

If the boundary conditions are other than u (and v) specified on the boundary or natural boundary conditions, then the functional again requires modification. This is illustrated in the case of a problem in two space variables with second derivatives present in the integral. Consider the problem of finding the function $u(x, y)$ which gives a stationary value to the functional

$$I(u) = \iint_R F(x, y, u, u_x, u_y, u_{xx}, u_{yy})\,dxdy$$

$$+ \int_{\partial R} G(x, y, u, u_s, u_{ss}, u_n)\,ds$$

where $\partial/\partial s$ and $\partial/\partial n$ are partial differential operators in the directions of the tangent and normal to the curve ∂R. The necessary conditions for $I(u)$ to have a minimum value are the Euler–Lagrange equation

$$F - \frac{\partial}{\partial x}F_{u_x} - \frac{\partial}{\partial y}F_{u_y} + \frac{\partial^2}{\partial x^2}F_{u_{xx}} + \frac{\partial^2}{\partial x\partial y}F_{u_{yy}} + \frac{\partial^2}{\partial y^2}F_{u_{yy}} = 0$$

together with the boundary conditions

$$\left[F_{u_x} - \frac{\partial}{\partial x}F_{u_{xx}}\right]\frac{dy}{ds} - \left[F_{u_y} - \frac{\partial}{\partial y}F_{u_{yy}}\right]\frac{dx}{ds}$$

$$- \left\{\frac{\partial}{\partial s}(F_{u_{xx}} - F_{u_{yy}})\right\}\frac{dx}{ds}\frac{dy}{ds} + \frac{1}{2}\left\{\frac{\partial}{\partial s}F_{u_{xy}}\left\{\left(\frac{dx}{ds}\right)^2 - \left(\frac{dy}{ds}\right)^2\right\}\right\}$$

$$+ \frac{1}{2}\left\{\left(\frac{\partial}{\partial x}F_{u_{xy}}\right)\frac{dx}{ds} - \left(\frac{\partial}{\partial y}F_{u_{xy}}\right)\frac{dy}{ds}\right\} + G_u - \frac{\partial}{\partial s}G_{u_s} + \frac{\partial^2}{\partial s^2}G_{u_{ss}} = 0 \qquad (2.5)$$

and

$$\frac{\partial G}{\partial u_n} + \frac{\partial F}{\partial u_{xx}}\left(\frac{dy}{ds}\right)^2 + \frac{\partial F}{\partial u_{yy}}\left(\frac{dx}{ds}\right)^2 + \frac{\partial F}{\partial u_{xy}}\frac{dx}{ds}\frac{dy}{ds} = 0. \qquad (2.6)$$

The function G is chosen so that the boundary conditions given by (2.5) and (2.6) correspond to the natural boundary conditions of the problem.

2.4 Time-dependent Variational Principles

The most basic and important time-dependent variational principle is *Hamilton's* principle from which can be deduced the fundamental equations of a large number of physical phenomena. Hamilton's principle states that the motion of a system from time t_0 to time t_1 is such that the time integral of the difference between the kinetic and potential energies is stationary for the true path. This can be expressed in mathematical terms by defining the integral in terms of the Lagrangian L as

$$I = \int_{t_0}^{t_1} L \, dt = \int_{t_0}^{t_1} (T - V) \, dt$$

where T, V are the respective kinetic and potential energies of the system, and stating that I is made stationary by the actual motion compared with neighbouring virtual motions. For a system with n generalized coordinates, $q_1, q_2, \ldots, q_n$, the associated Euler–Lagrange equations are

$$\frac{d}{dt}\left(\frac{\partial T}{\partial \dot{q}_r}\right) - \frac{\partial}{\partial q_r}(T - V) = 0 \quad (r = 1, 2, \ldots, n).$$

These are usually referred to as Lagrange's equations of motion for the system.

As a simple example of the foregoing theory applied to a continuum, we consider the case of a flexible string under constant tension τ. The string which is fixed at the ends executes small vibrations about the position of stable equilibrium, which is the interval $0 \leqslant x \leqslant 1$ of the x-axis. If $u(x, t)$ is the displacement perpendicular to the x-axis of a point on the string, then

$$T = \tfrac{1}{2}\rho \int_0^1 \left(\frac{\partial u}{\partial t}\right)^2 dx \quad \text{and} \quad V = \tfrac{1}{2}\rho \int_0^1 c^2 \left(\frac{\partial u}{\partial x}\right)^2 dx$$

where ρ is the density of the string and $c^2 = \tau/\rho$. The Euler–Lagrange equation is

$$\frac{\partial^2 u}{\partial t^2} = c^2 \frac{\partial^2 u}{\partial x^2}$$

which is the wave equation of the string. Thus the wave equation for the string is equivalent to the requirement that the difference between the total kinetic and potential energies of the string be as small as possible, on average, subject to the initial and boundary conditions of the problem. Other examples of the theory of this section are the vibrating rod, membrane, and plate (Courant and Hilbert, 1953, pp. 244–251).

Time-dependent variational principles are not used extensively at present for the numerical solution of evolutionary problems. These are usually solved by semi-discrete Galerkin methods (see Chapter 5).

We now turn our attention to time-dependent *dissipative* systems, and show how variational principles can be constructed for such systems. The method adopted is to introduce an adjoint system with negative friction. The energy lost by the dissipative system is gained by the adjoint system and so the total energy of the two systems is conserved. As an example consider the one-dimensional oscillator with friction. Its equation of motion is

$$\ddot{x} + k\dot{x} + n^2 x = 0 \quad (k > 0).$$

It is required to find a variational principle which has this equation as its Euler–Lagrange equation. This is impossible, but if we introduce the adjoint oscillator (represented by the coordinate x^*) with negative friction, its equation of motion is

$$\ddot{x}^* - k\dot{x}^* + n^2 x^* = 0.$$

The purely formal Lagrangian

$$L = \dot{x}\dot{x}^* - \tfrac{1}{2}k(x^*\dot{x} - x\dot{x}^*) - n^2 x x^*$$

will be seen to give the above two equations of motion as its Euler–Lagrange equations.

Another important example of a dissipative system is the heat diffusion problem. The governing equation for such a problem in one dimension is

$$\frac{\partial u}{\partial t} = \frac{\partial^2 u}{\partial x^2}$$

and we introduce the adjoint problem which is governed by the equation

$$-\frac{\partial u^*}{\partial t} = \frac{\partial^2 u^*}{\partial x^2}.$$

The formal Lagrangian in this case is

$$L = -\frac{\partial u}{\partial x}\frac{\partial u^*}{\partial x} - \frac{1}{2}\left(u^*\frac{\partial u}{\partial t} - u\frac{\partial u^*}{\partial t}\right)$$

which gives the above two equations as its Euler–Lagrange equations.

2.5 Dual Variational Principles

So far our variational principles have been one-sided, i.e. the approximate solution always lies above or below the theoretical solution of the variational principle. It is often possible, however, to construct two variational principles for a problem, where the same quantity d is a minimum and maximum with respect to the two principles. If d^u and d^l are approximate solutions of the minimum and maximum principles respectively, then

$$d^l \leqslant d \leqslant d^u$$

and so we have a practical method of bounding d. It is to be hoped that the quantity d is of physical significance. For further details the interested reader is referred to Barnsley and Robinson (1977), Sewell and Noble (1978), and Arthurs (1980).

Some examples will now be given of problems for which dual variational principles can be constructed.

(1) *The classical Dirichlet problem*   Here the functional

$$I(u) = \iint_R \tfrac{1}{2}[u_x^2 + u_y^2]\,\mathrm{d}x\mathrm{d}y$$

is minimized with respect to continuous functions $u(x, y)$ which have piecewise continuous derivatives in the region R and take prescribed values $u = f(s)$ on ∂R, where s is the arc length of ∂R, the boundary of R. The complementary or dual problem is the functional

$$J(v) = -\iint_R \tfrac{1}{2}\{v_x^2 + v_y^2\}\,\mathrm{d}x\mathrm{d}y - \int_{\partial R} vf'(s)\,\mathrm{d}s$$

maximized with respect to continuous functions $v(x, y)$ which have piecewise continuous derivatives in R and satisfy natural boundary conditions on ∂R. In this example

$$\min_u I(u) = \max_v J(v) = d.$$

(2) *Small displacement theory of elasticity* (Washizu, 1968). Consider an isotropic body in three-dimensional space occupying a region R enclosed by surface ∂R. The components of the body forces per unit volume are (X, Y, Z). The surface of the body is divided into two parts, S_f where the boundary conditions consist of external forces $(\bar{X}, \bar{Y}, \bar{Z})$ per unit area, and S_d over which the displacements $(\bar{u}, \bar{v}, \bar{w})$ are given. We have $\partial R = S_f + S_d$. Now the total potential energy is given by

$$I_p = \iiint_R W(\varepsilon_x, \varepsilon_y, \varepsilon_z, \gamma_{yz}, \gamma_{zx}, \gamma_{xy})\,\mathrm{d}\mathbf{x}$$

$$- \iiint_R (Xu + Yv + Zw)\,\mathrm{d}\mathbf{x} - \int_{S_f} (\bar{X}u + \bar{Y}v + \bar{Z}w)\,\mathrm{d}s$$

where

$$W = \frac{Ev}{2(1 + v)(1 - 2v)}(u_x + v_y + w_z)^2 + \frac{E}{2(1 + v)}(u_x^2 + u_y^2 + w_z^2)$$

$$+ \frac{E}{4(1 + v)}[(v_z + w_y)^2 + (w_x + u_z)^2 + (u_y + v_x)^2]$$

and where E and v are Young's modulus and Poisson's ratio, respectively, for the material. If the body forces and the surface forces are kept unchanged during variation, I_p is a minimum due to the actual displacements. This is the principle of *minimum potential energy*.

The complementary energy is given by

$$I_c = \iiint_R \varphi(\sigma_x, \sigma_y, \sigma_z, \tau_{yz}, \tau_{zx}, \tau_{xy}) \, dx - \int_{S_d} (X\bar{u} + Y\bar{v} + Z\bar{w}) \, ds$$

where

$$\varphi = \frac{1}{2E} [(\sigma_x + \sigma_y + \sigma_z)^2$$

$$+ 2(1 + v)(\tau_{yz}^2 + \tau_{zx}^2 + \tau_{xy}^2 - \sigma_y \sigma_z - \sigma_z \sigma_x - \sigma_x \sigma_y)].$$

If the surface displacements are kept unchanged during variation, I_c is a minimum due to the actual stresses. This is the principle of *minimum complementary energy*.

The quantity which can be bounded conveniently by these two principles is the direct influence coefficient or generalized displacement (Pian, 1970).

(3) *Compressible flow* (Sewell, 1969)   The respective volume integrands which appear in the dual variational principles are the pressure p and the quantity $p + \rho v^2$, where ρ is the density and v is the velocity of the fluid. Here

$$p = p(v_i, h, \eta)$$

where $v_i(i = 1, 2, 3)$ are the velocity components and h and η are the total energy and entropy per unit mass respectively. The results

$$\frac{\partial p}{\partial v_i} = -Q_i \quad (i = 1, 2, 3) \quad \frac{\partial p}{\partial h} = \rho \quad \frac{\partial p}{\partial \eta} = -\rho T$$

follow, where $Q_i = \rho v_i (i = 1, 2, 3)$ with ρ the density and T the temperature. The function

$$P = P(Q_i, h, \eta) = \sum_{i=1}^{3} Q_i v_i + p$$

is introduced, where

$$\frac{\partial P}{\partial Q_i} = v_i \quad (i = 1, 2, 3) \quad \frac{\partial P}{\partial h} = \rho \quad \frac{\partial P}{\partial \eta} = -\rho T.$$

The dual variational principles involving p and P respectively can be strengthened to extremum principles for particular types of compressible flow.

(4) *Dissipative systems*   Dual extremum principles for the heat equation and similar equations can be obtained by imbedding the initial-value problem in a

two-point boundary value problem for a system which couples the original equation with its adjoint. Details of such procedures can be found in Collins (1977), Herrera and Sewell (1978) and Noble and Sewell (1972).

Exercise 2.4 An incompressible inviscid flow is parallel to the x-axis. Calculate $I(u)$ and $J(v)$ for this problem where R is the square region $0 \leqslant x, y \leqslant 1$, and show that the extreme values coincide. (*Hint* The functions u and v are the stream function and potential respectively for the flow.)

Exercise 2.5 Show that the necessary conditions for $J(v)$ to have a maximum consist of the Euler–Lagrange equation

$$v_{xx} + v_{yy} = 0$$

together with the natural boundary condition

$$v_y \frac{dx}{ds} - v_x \frac{dy}{ds} = f'(s).$$

Exercise 2.6 Show that $W = \varphi$ if the following linear stress–strain relations hold:

$$E\varepsilon_x = \sigma_x - v(\sigma_y + \sigma_z) \quad \tau_{yz} = \frac{E}{2(1 + v)}\gamma_{yz}$$

$$E\varepsilon_y = \sigma_y - v(\sigma_x + \sigma_z) \quad \tau_{zx} = \frac{E}{2(1 + v)}\gamma_{zx}$$

$$E\varepsilon_z = \sigma_z - v(\sigma_x + \sigma_y) \quad \tau_{xy} = \frac{E}{2(1 + v)}\gamma_{xy}.$$

Exercise 2.7 Show that the necessary conditions for the potential energy

$$I_p = \int\int_R \left[\frac{Ev}{2(1 + v)(1 - 2v)}(u_x + v_y)^2 + \frac{E}{2(1 + v)}(u_x^2 + v_y^2) + \frac{E}{4(1 + v)}(u_y + v_x)^2 \right] dxdy$$

to have a minimum are the Euler–Lagrange equations

$$(2 - 2v)u_{xx} + (1 - 2v)u_{yy} + v_{xy} = 0$$

and

$$(2 - 2v)v_{yy} + (1 - 2v)v_{xx} + u_{xy} = 0$$

in the region R together with the boundary conditions

$$(2 - 2v)u_x \cos \alpha + (1 - 2v)u_y \sin \alpha + (1 - 2v)v_x \sin \alpha + 2vv_y \cos \alpha = 0$$

and

$$2vu_x \sin \alpha + (1 - 2v)u_y \cos \alpha + (1 - 2v)v_x \cos \alpha + (2 - 2v)v_y \sin \alpha = 0$$

on ∂R (see Figure 2.1). This is *plane strain* applicable to thick solids with no variation in the z-direction (cf. *plane stress* in Section 2.2).

3

Basis Functions

In Section 1.1, elementary basis functions were constructed for rectangular and polygonal regions, where the former was divided up into a number of rectangular elements and the latter into a number of triangular elements. In this chapter, a study is made of the construction of basis functions for a variety of element shapes in two and three dimensions.

3.1 The Triangle

3.1.1 Lagrange interpolation

The triangle or two-dimensional simplex is probably the most widely used finite element. One reason for this is that arbitrary regions in two dimensions can be approximated by polygons, which can always be divided up into a finite number of triangles. In addition, the complete mth-order polynomial

$$\Pi_m(x, y) = \sum_{k+l=0}^{m} \alpha_{kl} x^k y^l \tag{3.1}$$

can be used to interpolate a function, say $U(x, y)$, at $\frac{1}{2}(m + 1)(m + 2)$ symmetrically placed nodes in a triangle. The first three cases of this general representation for the triangle $P_1 P_2 P_3$, with the coordinates of the vertices being (x_1, y_1), (x_2, y_2), and (x_3, y_3) respectively, are:

(1) *The linear case* $(m = 1)$ Here the polynomial is

$$\Pi_1(x, y) = \alpha_1 + \alpha_2 x + \alpha_3 y$$

$$= \sum_{j=1}^{3} U_j p_j^{(1)}(x, y)$$

41

where $U_j (j = 1, 2, 3)$ are the values of $U(x, y)$ at the vertices P_j and

$$p_j^{(1)}(x, y) = \frac{1}{C_{jkl}} (\tau_{kl} + \eta_{kl} x - \xi_{kl} y)$$

$$= \frac{D_{kl}}{C_{jkl}} \tag{3.2}$$

where

$$\tau_{kl} = x_k y_l - y_k x_l \quad \xi_{kl} = x_k - x_l \quad \eta_{kl} = y_k - y_l$$

and

$$D_{kl} = \det \begin{bmatrix} 1 & x & y \\ 1 & x_k & y_k \\ 1 & x_l & y_l \end{bmatrix}$$

with (j, k, l) any permutation of $(1, 2, 3)$, and the modulus of

$$C_{jkl} = \det \begin{bmatrix} 1 & x_j & y_j \\ 1 & x_k & y_k \\ 1 & x_l & y_l \end{bmatrix}$$

is twice the area of the triangle $P_1 P_2 P_3$. It is easily seen that

$$p_j^{(1)}(x_k, y_k) = \begin{cases} 1 & (j = k) \\ 0 & (j \neq k) \end{cases} \quad (1 \leqslant j, k \leqslant 3).$$

(2) *The quadratic case* ($m = 2$)   The polynomial is now

$$\Pi_2(x, y) = \sum_{j=1}^{6} U_j p_j^{(2)}(x, y) \tag{3.3}$$

where $U_j (j = 1, \ldots, 6)$ are the values of $U(x, y)$ at the vertices $P_j (j = 1, 2, 3)$ together with the values at the midpoints $P_j (j = 4, 5, 6)$ of the sides $P_1 P_2, P_2 P_3$, and $P_3 P_1$ respectively. The functions $p_j^{(2)}(x, y)$ ($j = 1, 2, \ldots, 6$) are given by

$$p_1^{(2)}(x, y) = p_1^{(1)}(2p_1^{(1)} - 1)$$

with $p_2^{(2)}(x, y)$ and $p_3^{(2)}(x, y)$ similarly, and

$$p_4^{(2)}(x, y) = 4p_1^{(1)} p_2^{(1)}$$

with $p_5^{(2)}(x, y)$ and $p_6^{(2)}(x, y)$ similarly. Again it follows that

$$p_j^{(2)}(x_k, y_k) = \begin{cases} 1 & (j = k) \\ 0 & (j \neq k) \end{cases} \quad (1 \leqslant j, k \leqslant 6).$$

It is particularly satisfactory that the basis functions $p_j^{(2)}(x, y)$ ($j = 1, \ldots, 6$) can be expressed in terms of the basis function $p_j^{(1)}(x, y)$ ($j = 1, 2, 3$). This is true for nearly all the basis functions we shall consider and therefore in order to simplify the formulae we shall denote $p_j^{(1)}$ simply by $p_j(j = 1, 2, 3)$.

(3) *The cubic case* $(m = 3)$ The polynomial is

$$\Pi_3(x, y) = \sum_{j=1}^{10} U_j p_j^{(3)}(x, y) \tag{3.4}$$

where $U_j(j = 1, 2, 3)$ are the values of $U(x, y)$ at the vertices P_1, P_2, P_3, $U_j(j = 4, 5, \ldots, 9)$ are the values at the points of trisection of the sides, and U_{10} is the value of $U(x, y)$ at the centroid of the triangle. The basis functions are given by

$$p_1^{(3)}(x, y) = \tfrac{1}{2}p_1(3p_1 - 1)(3p_1 - 2)$$

with $p_2^{(3)}(x, y)$ and $p_3^{(3)}(x, y)$ similarly,

$$p_4^{(3)}(x, y) = \tfrac{9}{2}p_1 p_2(3p_1 - 1)$$
$$p_5^{(3)}(x, y) = \tfrac{9}{2}p_1 p_2(3p_2 - 1)$$

with $p_6^{(3)}, \ldots, p_9^{(3)}$ similarly, and

$$p_{10}^{(3)}(x, y) = 27 p_1 p_2 p_3.$$

The tenth parameter can be eliminated by using the linear relation

$$U_{10} = \frac{1}{4} \sum_{j=4}^{9} U_j - \frac{1}{6} \sum_{j=1}^{3} U_j$$

to yield a function that will still recover quadratics exactly (Ciarlet and Raviart, 1972a); this is an example of the so-called *elimination of internal parameters*. The triangles for the cases $m = 1, 2, 3$, are shown in Figure 3.1.

We now turn to the general case where the complete mth-order polynomial is

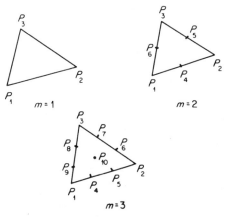

Figure 3.1

given by (3.1). This polynomial has $\frac{1}{2}(m+1)(m+2)$ coefficients which can be chosen so that the polynomial interpolates $U(x,y)$ at the $\frac{1}{2}(m+1)(m+2)$ symmetrically placed points on the triangle $P_1P_2P_3$ whose coordinates are given by

$$\sum_{l=1}^{3} \frac{\beta_l x_l}{m}, \quad \sum_{l=1}^{3} \frac{\beta_l y_l}{m} \tag{3.5}$$

where β_1, β_2 and β_3 are integers satisfying $0 \leqslant \beta_k \leqslant m (k = 1, 2, 3)$ and $\beta_1 + \beta_2 + \beta_3 = m$. These points include the three vertices of the triangle $P_1P_2P_3$. The remaining points are obtained geometrically by dividing each side of the triangle into m equal parts and joining the points of subdivision by lines parallel to the sides of the triangle. This subdivides the triangle into m^2 congruent triangles whose vertices are the $\frac{1}{2}(m+1)(m+2)$ points described by (3.5). If U_j denotes the value of $U(x,y)$ at a point given by (3.5), the interpolating polynomial of degree m can be expressed as

$$U(x,y) = \sum_{j=1}^{\frac{1}{2}(m+1)(m+2)} U_j p_j^{(m)}(x,y)$$

where the summation is over all $\frac{1}{2}(m+1)(m+2)$ points, and $p_j^{(m)}(x,y)$ is a polynomial basis function of degree m taking the value unity at the point associated with the triple $(\beta_1, \beta_2, \beta_3)$ and the value zero at every other point.

Exercise 3.1 Verify that

$$U(x,y) = \sum_{j=1}^{9} U_j \bar{p}_j^{(3)}(x,y)$$

recovers quadratic polynomials exactly if

$$\bar{p}_j^{(3)} = p_j^{(3)} + \alpha_j p_{10}^{(3)} \quad (j = 1, \ldots, 9) \tag{3.6}$$

with

$$\alpha_j = \begin{cases} -\frac{1}{6} & (j = 1, 2, 3) \\ \frac{1}{4} & (j = 4, \ldots, 9). \end{cases}$$

Exercise 3.2 Show that

$$p_j^{(3)}(x_k, y_k) = \begin{cases} 1 & (k = j) \\ 0 & (k \neq j) \end{cases} \quad (1 \leqslant j, k \leqslant 10).$$

3.1.2 The standard triangle

It should be noted from (3.2) that

(i) $\sum_{j=1}^{3} p_j(x,y) = 1,$

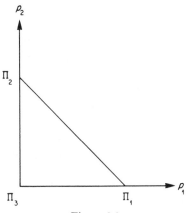

Figure 3.2

(ii) the linear equations $p_j(x, y) = 0\,(j = 1, 2, 3)$ represent the triangle sides $P_2 P_3, P_3 P_1$, and $P_1 P_2$ respectively, and

(iii) $p_j(x, y) = 1\,(j = 1, 2, 3)$ at the vertices P_1, P_2, and P_3 respectively.

Alternatively, the triangle $P_1 P_2 P_3$ in the (x, y) plane is transformed into the *standard triangle* $\Pi_1 \Pi_2 \Pi_3$ in the (p_1, p_2) plane by the transformation formulae (3.2), where $\Pi_1 = (1, 0)$, $\Pi_2 = (0, 1)$, and $\Pi_3 = (0, 0)$ (see Figure 3.2). The inverse transformation from the (p_1, p_2) plane to the (x, y) plane is given by

$$x = x_3 + \xi_{13} p_1 + \xi_{23} p_2$$

and

$$y = y_3 + \eta_{13} p_1 + \eta_{23} p_2. \tag{3.7}$$

Since all triangles in a triangular network in the (x, y) plane can be transformed into this standard triangle, it is very convenient to work in terms of the standard triangle, and at an appropriate point to transfer the result back to a particular triangle in the (x, y) plane through the linear transformation (3.7). This procedure will be used repeatedly when triangular elements are involved both in this chapter and in Chapter 6.

A simple geometrical procedure for constructing basis functions obtained from Lagrange interpolation is now illustrated with respect to the quadratic and cubic cases of the triangle. In the former, for example, the basis function $p_1^{(2)}(x, y)$ must take the value zero at the nodes $P_j(j = 2, 3, \ldots, 6)$ and the value unity at the node P_1. The line $p_1 = 0$ passes through the points P_2, P_3, and P_5, and the line $p_1 = \frac{1}{2}$ passes through the points P_4 and P_6, and so the basis function $p_1^{(2)}(x, y) = p_1(2p_1 - 1)$ is obtained. The functions $p_2^{(2)}(x, y)$ and $p_3^{(2)}(x, y)$ are similarly obtained. The basis function at a midside node, say P_4, is obtained from the lines

$p_1 = 0$ and $p_2 = 0$, which pass through the points P_2, P_3, P_5, and P_1, P_3, P_6 respectively. The required basis function, normalized at P_4, is $p_4^{(2)}(x, y) = 4p_1 p_2$. The functions $p_5^{(2)}(x, y)$ and $p_6^{(2)}(x, y)$ are similarly obtained. The cubic case follows in an analogous manner, provided the internal node has not been eliminated.

In the general case, the function interpolates the values of $U(x, y)$ at $\frac{1}{2}(m + 1) \times (m + 2)$ points on the triangle. On a side of the triangle, this function reduces to a polynomial of degree m in the variable s which is measured along the side of the triangle. The polynomial interpolates $U(x, y)$ at $(m + 1)$ points on the side of the triangle, and so is unique. Also in a triangular network, each side (internal) belongs to two triangles. If the function in each triangle interpolates at $\frac{1}{2}(m + 1) \times (m + 2)$ symmetrically placed points, it will reduce to the unique polynomial of degree m in s on the common side. This means that the interpolating function over the complete triangular network is continuous along internal sides of the network and so has C^0 continuity over the polygonal region.

Exercise 3.3   Verify that (3.4) can be obtained using the geometric method.

3.1.3 Hermite interpolation

As an alternative to interpolating the function $U(x, y)$ at a large number of points symmetrically placed on the triangle, it is possible to interpolate $U(x, y)$ and some of its derivatives at a smaller number of points.

A class of polynomials suited to this particular task consists of the complete polynomials $G_v(x, y)$ of odd degree $2v + 1$ ($v = 1, 2, 3, \ldots$), which are determined by the values

$$D^i G_v(P_j) \quad (|i| \leqslant v; j = 1, 2, 3)$$

and

$$D^i G_v(P_4) \quad (|i| \leqslant v - 1). \tag{3.8}$$

Here P_1, P_2, P_3 are the vertices of the triangle, and P_4 is the centroid; $i = (i_1, i_2)$ where i_1, i_2 are nonnegative integers, $|i| = i_1 + i_2$, and

$$D^i G = \frac{\partial^{|i|} G}{\partial x^{i_1} \partial y^{i_2}}.$$

This is an example of the *multi-index notation* for derivatives. The first two cases of this general representation are:

(1) *The cubic case* ($v = 1$)   Here the complete cubic polynomial has ten coefficients which are uniquely determined by matching the function values and the first-order partial derivatives at the vertices and the function value at the centroid. Thus in this case we may write the polynomial $\Pi_3(x, y)$, which is given in

(3.4), in the form

$$G_1(x, y) = \sum_{j=1}^{4} U_j q_j^{(3)}(x, y) + \sum_{j=1}^{3} \left[\left(\frac{\partial U}{\partial x} \right)_j r_j^{(3)}(x, y) + \left(\frac{\partial U}{\partial y} \right)_j s_j^{(3)}(x, y) \right] \quad (3.9)$$

where

$$q_j^{(3)}(x, y) = p_j(3p_j - 2p_j^2 - 7p_k p_l)$$

with (j, k, l) any cyclic permutation of $(1, 2, 3)$ and

$$q_4^{(3)}(x, y) = 27 p_1 p_2 p_3.$$

Also

$$r_j^{(3)}(x, y) = p_j[\xi_{jk} p_k(p_l - p_j) + \xi_{jl} p_l(p_k - p_j)]$$

and $s_j^{(3)}(x, y)$ is obtained from $r_j^{(3)}(x, y)$ by replacing ξ by η. This element is in common use in the finite element method.

(2) *The quintic case* $(v = 2)$ This time the complete quintic polynomial has twenty-one coefficients which are uniquely determined by matching the function values and the first- and second-order partial derivatives at the vertices, and the function value and the first-order partial derivatives at the centroid. The element is of little practical use and will not be considered further.

As for cubic Lagrange interpolation it is possible to eliminate the value of $U(x, y)$ at the centroid and still recover quadratic polynomials exactly. In this case the replacement is

$$U_4 = \frac{1}{3} \sum_{j=1}^{3} U_j + \frac{1}{18} \sum^{(1)} \left\{ \left(\frac{\partial U}{\partial x} \right)_j (\xi_{kj} + \xi_{lj}) + \left(\frac{\partial U}{\partial x} \right)_j (\eta_{kj} + \eta_{lj}) \right\}$$

where $\sum^{(1)}$ denotes the summation over (j, k, l) for all cyclic permutations of $(1, 2, 3)$.

The interpolating polynomial now becomes

$$G_1^*(x, y) = \sum_{j=1}^{3} \left[U_j q_j^{(3)*}(x, y) + \left(\frac{\partial U}{\partial x} \right)_j r_j^{(3)*}(x, y) + \left(\frac{\partial U}{\partial y} \right)_j s_j^{(3)*}(x, y) \right] \quad (3.10)$$

where

$$q_j^{(3)*}(x, y) = p_j(3p_j - 2p_j^2 + 2p_k p_l) \quad (3.10a)$$

and

$$r_j^{(3)*}(x, y) = p_j^2(p_l \xi_{lj} + p_k \xi_{kj}) + \tfrac{1}{2} p_j p_k p_l(\xi_{lj} + \xi_{kj}) \quad (3.10b)$$

with (j, k, l) any permutation of $(1, 2, 3)$. The functions $s_j^{(3)*}$ are obtained from (3.10b) by replacing ξ by η.

Birkhoff (1971) introduced a triangular element which involves the twelve-parameter family of all quartic polynomials which are cubic along any parallel to any side of a triangle. With respect to the standard triangle, such a family is

$$U(p,q) = \sum_{j+k \leqslant 4} \alpha_{jk} p_1^j p_2^k \tag{3.11}$$

with

$$\alpha_{31} + \alpha_{13} = \alpha_{22} \quad \text{and} \quad \alpha_{40} = \alpha_{04} = 0.$$

The polynomial (3.11) is called a *tricubic* polynomial. This polynomial uniquely interpolates values of U, $\partial U/\partial p_1$, $\partial U/\partial p_2$ and $\partial^2 U/\partial r \partial s$ at each vertex, where $\partial^2 U/\partial r \partial s$ is a cross-derivative determined at each vertex with r and s in directions parallel to the adjacent sides. The unique interpolating function takes the form

$$\sum_{j=1}^{3} \left[\left\{ \sum_{|i| \leqslant 1} \mathbf{D}^i U_j \varphi_j^i(p_1, p_2) \right\} + \left(\frac{\partial^2 U}{\partial r \partial s} \right)_j \hat{\phi}_j(p_1, p_2) \right] \tag{3.12}$$

where the suffix denotes the value of the quantity at the vertex Π_j of the standard triangle (Figure 3.2). The coefficients are given by

$$\varphi_j^{(0,0)} = p_j^2(3 - 2p_j + 6p_k p_l)$$

$$\varphi_j^{(1,0)} = \begin{cases} p_1^2(p_1 - 1 - 4p_2 p_3) & (j = 1) \\ p_2^2 p_1(1 + 2p_3) & (j = 2) \\ p_3^2 p_1(1 + 2p_2) & (j = 3) \end{cases}$$

$$\varphi_j^{(0,1)} = \begin{cases} p_1^2 p_2(1 + 2p_3) & (j = 1) \\ p_2^2(p_2 - 1 - 4p_1 p_3) & (j = 2) \\ p_3^2 p_2(1 + 2p_1) & (j = 3) \end{cases}$$

and

$$\hat{\phi}_j = 2p_j^2 p_k p_l$$

where (j, k, l) is any permutation of $(1, 2, 3)$. Note that $\mathbf{D}^i$ in this case represents derivatives in the (p_1, p_2) plane. The cross-derivatives $(\partial^2 U/\partial r \partial s)_j (j = 1, 2, 3)$ are given by $-\partial^2 U/\partial p_1 \partial Q, \partial^2 U/\partial p_2 \partial Q$ and $\partial^2 U/\partial p_1 \partial p_2$ respectively, where $Q = p_2 - p_1$. The unique tricubic interpolating polynomial in a particular triangle in the (x, y) plane is obtained from (3.12) by using the linear transformation formulae (3.2).

Exercise 3.4 Show that on a side of the triangle the function given by (3.9) reduces to a polynomial of degree 3 in the variable s which is measured along the side of the triangle. Show also that this polynomial is uniquely determined by the values of the function and its first-order partial derivatives at the two vertices which are the endpoints of the side. Hence show that this element gives an interpolating function which has C^0 continuity over a triangular network.

Exercise 3.5 Verify that quadratic functions are interpolated exactly by the reduced cubic interpolant.

Exercise 3.6 By using the linear transformation formulae (3.7), show that (3.10) becomes

$$G_1^*(p_1, p_2) = \sum_{j=1}^{3} \left[U_j q_j^{(3)^*} + \left(\frac{\partial U}{\partial p_1}\right)_j r_j^{(3)^*} + \left(\frac{\partial U}{\partial p_2}\right)_j s_j^{(3)^*} \right]$$

in the variables of the standard triangle, where

$$\frac{\partial U}{\partial p_j} = \xi_{j3}\frac{\partial U}{\partial x} + \eta_{j3}\frac{\partial U}{\partial y} \quad (j = 1, 2)$$

and the functions $q_j^{(3)^*}$, $r_j^{(3)^*}$ and $s_j^{(3)^*}$ $(j = 1, 2, 3)$ are given by

$$q_j^{(3)^*} = p_j^2(3 - 2p_j) + 2p_1 p_2 p_3$$

$$r_j^{(3)^*} = \begin{cases} p_1^2(p_1 - 1) - p_1 p_2 p_3 & (j = 1) \\ p_j^2 p_1 + \frac{1}{2}p_1 p_2 p_3 & (j = 2, 3) \end{cases}$$

and

$$s_j^{(3)^*} = \begin{cases} p_2^2(p_2 - 1) - p_1 p_2 p_3 & (j = 2) \\ p_j^2 p_2 + \frac{1}{2}p_1 p_2 p_3 & (j = 1, 3). \end{cases}$$

You may use the results

$$\xi_{kl}\eta_{lj} - \xi_{lj}\eta_{kl} = C_{jkl} \quad \text{(see p. 42)}$$

where (j, k, l) is any permutation of $(1, 2, 3)$.

Exercise 3.7 Show that the tricubic polynomials produce a C^0 approximating function over a network of triangular elements.

3.1.4 C^1 Approximating functions

A triangular element is now introduced which involves the complete family of quintic polynomials. In the plane of the standard triangle, the complete quintic is

$$U(p_1, p_2) = \sum_{j+k \leqslant 5} \alpha_{jk} p_1^j p_2^k. \tag{3.13}$$

The coefficients α_{jk} can be determined in terms of

$$D^i U(P_j) \quad (|i| \leqslant 2; j = 1, 2, 3)$$

and

$$\frac{\partial U}{\partial n}(P_j) \quad (j = 4, 5, 6)$$

where $P_j(j = 1, 2, 3)$ are the vertices and $P_j(j = 4, 5, 6)$ are the side midpoints. The function $U(p_1, p_2)$ given by (3.13) reduces to a quintic in s along each side of the triangle, which is uniquely determined by the parameters at the vertices which provide six boundary conditions, viz. $U, \partial U/\partial s, \partial^2 U/\partial s^2$ at the endpoints of each side. The normal derivative to each side, $\partial U/\partial n$, where n is p_2, $p_1 + p_2$ and p_1

respectively is a quartic in s, and is uniquely determined by the parameters $\partial U/\partial n$ and $\partial^2 U/\partial n\partial s$ at the endpoints of each side, together with $\partial U/\partial n$ at midside points. It has thus been shown that complete quintic polynomials produce an approximating function over a network of triangular elements, which has continuity of displacement and gradient over the complete region. Such an interpolating function is said to be C^1 over the region.

In fact the parameters corresponding to the normal derivatives at the midside points can be eliminated without destroying the C^1 continuity over the triangular network. This is accomplished by imposing a cubic variation of the normal derivative along each side, which is equivalent to requiring that in (3.13)

$$\alpha_{41} = \alpha_{14} = 0$$

and

$$5\alpha_{50} + \alpha_{32} + \alpha_{23} + 5\alpha_{05} = 0.$$

This reduces (3.13) to an eighteen-parameter family of polynomials, and the unique interpolating function is given by

$$U(p_1, p_2) = \sum_{j=1}^{3} \sum_{|i| \leqslant 2} D^i U_j \varphi_j^i(p_1, p_2) \tag{3.14}$$

where

$$\varphi_1^{(0,0)} = p_1^2(10p_1 - 15p_1^2 + 6p_1^3 + 15p_2^2 p_3)$$

$$\varphi_2^{(0,0)} = p_2^2(10p_2 - 15p_2^2 + 6p_2^3 + 15p_1^2 p_3)$$

$$\varphi_3^{(0,0)} = p_3^2(10p_3 - 15p_3^2 + 6p_3^3 + 30p_1 p_2(p_1 + p_2))$$

$$\varphi_1^{(1,0)} = p_1^2(-4p_1 + 7p_1^2 - 3p_1^3 - \tfrac{15}{2}p_2^2 p_3)$$

$$\varphi_2^{(1,0)} = p_1 p_2^2(3 - 2p_2 - \tfrac{3}{2}p_1 - \tfrac{3}{2}p_1^2 + \tfrac{3}{2}p_1 p_2)$$

$$\varphi_3^{(1,0)} = p_1 p_3^2(3 - 2p_3 - 3p_1^2 + 6p_1 p_2)$$

$$\varphi_1^{(0,1)} = p_1^2 p_2(3 - 2p_1 - \tfrac{3}{2}p_3 - \tfrac{3}{2}p_2^2 + \tfrac{3}{2}p_1 p_2)$$

$$\varphi_2^{(0,1)} = p_2^2(-4p_2 + 7p_2^2 - 3p_2^3 - \tfrac{15}{2}p_1^2 p_3)$$

$$\varphi_3^{(0,1)} = p_2 p_3^2(3 - 2p_3 - 3p_2^2 + 6p_1 p_2)$$

$$\varphi_1^{(1,1)} = p_1^2 p_2(-1 + p_1 + \tfrac{1}{2}p_2 + \tfrac{1}{2}p_2^2 - \tfrac{1}{2}p_1 p_2)$$

$$\varphi_2^{(1,1)} = p_1 p_2^2(-1 + \tfrac{1}{2}p_1 + p_2 + \tfrac{1}{2}p_1^2 - \tfrac{1}{2}p_1 p_2)$$

$$\varphi_3^{(1,1)} = p_1 p_2 p_3^2$$

$$\varphi_1^{(2,0)} = p_1^2(\tfrac{1}{2}p_1(1 - p_1)^2 + \tfrac{5}{4}p_2^2 p_3)$$

$$\varphi_2^{(2,0)} = \tfrac{1}{4}p_1^2 p_2^2 p_3 + \tfrac{1}{2}p_1^2 p_2^3$$

$$\varphi_3^{(2,0)} = \tfrac{1}{2}p_1^2 p_3^2(1 - p_1 + 2p_2)$$

$$\varphi_1^{(0,2)} = \tfrac{1}{4}p_1^2 p_2^2 p_3 + \tfrac{1}{2}p_1^3 p_2^2$$

$$\varphi_2^{(0,2)} = p_2^2(\tfrac{1}{2}p_2(1 - p_2)^2 + \tfrac{5}{4}p_1^2 p_3)$$

and

$$\varphi_3^{(0,2)} = \tfrac{1}{2}p_2^2 p_3^2(1 + 2p_1 - p_2).$$

3.2 The Rectangle

3.2.1 Lagrange interpolation

Rectangular type regions occur in many problems in physics and engineering, ensuring an important rôle in the finite element method for the rectangular element. In the latter the bivariate polynomial

$$\sum_{k=0}^{m}\sum_{l=0}^{m}\alpha_{kl}x^{k}y^{l}$$

is used to interpolate $U(x, y)$ at $(m + 1)^2$ symmetrically placed nodes. For convenience, we consider the rectangle in the form of the standard unit square $0 \leqslant p, q \leqslant 1$.

(1) *The bilinear case* $(m = 1)$ Here the interpolant is

$$U(p, q) = \sum_{j=1}^{4} \varphi_j^{(1)}(p, q)U_j \tag{3.15}$$

with

$$\varphi_1^{(1)} = pq \quad \varphi_2^{(1)} = (1 - p)q \quad \varphi_3^{(1)} = (1 - p)(1 - q) \quad \varphi_4^{(1)} = p(1 - q)$$

where the four corner points numbered 1, 2, 3, and 4 are located at the positions $(1, 1)$, $(0, 1)$, $(0, 0)$, and $(1, 0)$ in (p, q)-space respectively.

(2) *The biquadratic case* $(m = 2)$ This time

$$U(p, q) = \sum_{j=1}^{9} \varphi_j^{(2)}(p, q)U_j$$

with

$$\varphi_1^{(2)} = p(2p - 1)q(2q - 1) \quad \varphi_2^{(2)}, \varphi_3^{(2)}, \text{ and } \varphi_4^{(2)} \text{ similarly}$$
$$\varphi_5^{(2)} = 4(1 - p)pq(2q - 1) \quad \varphi_6^{(2)}, \varphi_7^{(2)}, \text{ and } \varphi_8^{(2)} \text{ similarly}$$

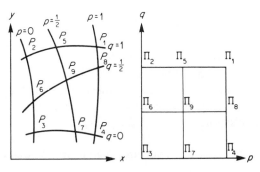

Figure 3.3

and

$$\varphi_9^{(2)} = 16p(1-p)q(1-q)$$

where the node numbering is as in Figure 3.3. The internal node 9 can be eliminated by using the relation

$$U_9 = \tfrac{1}{2} \sum_{j=5}^{8} U_j - \tfrac{1}{4} \sum_{j=1}^{4} U_j.$$

This yields a function which still interpolates up to quadratics in p and q exactly and we say that such an element has *quadratic precision*. There is no term in p^2q^2 in the approximation, which can now be written as

$$U(p, q) = \sum_{j=1}^{8} \varphi_j^{(2)^*}(p, q)U_j \qquad (3.16a)$$

where

and
$$
\left.
\begin{array}{ll}
\varphi_1^{(2)^*} = pq(2p + 2q - 3) & \varphi_j^{(2)^*}(j = 2,\ 3,\ 4) \quad \text{similarly} \\[2mm]
\varphi_5^{(2)^*} = 4pq(1 - p) & \varphi_j^{(2)^*}(j = 6,\ 7,\ 8) \quad \text{similarly.}
\end{array}
\right\} \qquad (3.16b)
$$

An alternative method of deriving interpolants which depend only on continuous or interpolated data on element boundaries is to use *blending function* interpolants (Gordon, 1971). For example, if $U \in C^{2,2}(R)$ where $R = [0, 1] \times [0, 1]$, the function

$$\tilde{U}(p, q) = (1 - p)U(0, q) + pU(1, q) + (1 - q)(p, 0) + qU(p, 1) - \tilde{\tilde{U}}(p, q) \qquad (3.17a)$$

where

$$\begin{aligned}
\tilde{\tilde{U}}(p, q) = {} & (1 - p)(1 - q)U(0, 0) + p(1 - q)U(1, 0) \\
& + (1 - p)qU(0, 1) + pqU(1, 1)
\end{aligned}$$

interpolates U exactly on the four sides of the square. In the present case,

$$\begin{bmatrix} U(p, 0) \\ U(p, 1) \end{bmatrix} = (1 - p)(1 - 2p)\begin{bmatrix} U_3 \\ U_2 \end{bmatrix} + 4p(1 - p)\begin{bmatrix} U_7 \\ U_5 \end{bmatrix} + p(2p - 1)\begin{bmatrix} U_4 \\ U_1 \end{bmatrix}$$

$$\begin{bmatrix} U(0, q) \\ U(1, q) \end{bmatrix} = (1 - q)(1 - 2q)\begin{bmatrix} U_3 \\ U_4 \end{bmatrix} + 4q(1 - q)\begin{bmatrix} U_6 \\ U_8 \end{bmatrix} + q(2q - 1)\begin{bmatrix} U_2 \\ U_1 \end{bmatrix}$$

$$(3.17b)$$

which on substitution into (3.17a) leads to (3.16a).

The geometrical method outlined for the triangle in the previous section applies also for the rectangular element, even after substitutions have taken place to remove any internal nodes.

(3) *The bicubic case* ($m = 3$)   The full bicubic approximation involves four

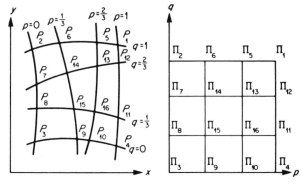

Figure 3.4

internal nodes (Figure 3.4) in addition to the four corner and eight side nodes (two on each side). The internal nodes can be eliminated to yield an approximation with no terms involving p^2q^2, p^2q^3, p^3q^2 and p^3q^3 that can be written as

$$U(p,q) = \sum_{j=1}^{12} \varphi_j^{(3)*}(p,q)U_j. \tag{3.18a}$$

If the quadratic interpolation (3.17b) is replaced by cubic interpolation, then the bilinear blending (3.17a) leads to

$$\varphi_1^{(3)*} = \tfrac{9}{2}(p^2 + q^2 - p - q + \tfrac{2}{9})pq$$

and similarly for $\varphi_j^{(3)*}$ $(j = 2, 3, 4)$,

$$\varphi_5^{(3)*} = \tfrac{9}{2}pq(1 - p)(3p - 1) \tag{3.18b}$$

and similarly for $\varphi_j^{(3)*}$ $(j = 6, \ldots, 12)$. Alternatively, basis functions can be written down directly using the geometrical procedure of Section 3.1.2 to give

$$\varphi_1^{(3)*} = \tfrac{1}{2}pq(3p + 3q - 4)(3p + 3q - 5)$$

and similarly for $\varphi_j^{(3)*}$ $(j = 2, 3, 4)$ $\tag{3.18c}$

$$\varphi_5^{(3)*} = \tfrac{9}{2}pq(1 - p)(3p - 1)$$

and similarly for $\varphi_j^{(3)*}$ $(j = 6, \ldots, 12)$.

Exercise 3.8 Verify that the geometric method leads to (3.16b) and (3.18c) whereas blending leads to (3.16b) and (3.18b).

3.2.2 Bicubic Hermite interpolation

In the previous section, bivariate functions were used to construct basis functions each of which was identically zero except for a region composed of four rectangular elements. Over the complete rectangular region the piecewise polynomial approximating function gave C^0 continuity. In this section we consider the bicubic polynomials written in the form

$$H_3(p,q) = \sum_{k=0}^{3} \sum_{l=0}^{3} \alpha_{kl} p^k q^l \qquad (3.19a)$$

over the unit square $0 \leqslant p, q \leqslant 1$. The coefficients $\alpha_{kl} (0 \leqslant k, l \leqslant 3)$ can be found uniquely in terms of the values of H_3, $\partial H_3/\partial p$, $\partial H_3/\partial q$ and $\partial^2 H_3/\partial p \partial q$ at the four corners such that

$$H_3(p,q) = \sum_{0 \leqslant k,l,L,M \leqslant 1} (D^{(L,M)} U)_{kl} \psi_k^{(L)}(p) \psi_l^{(M)}(q) \qquad (3.19b)$$

where

$$\psi_0^{(0)}(t) = (1-t)^2 (1+2t)$$
$$\psi_0^{(1)}(t) = (1-t)^2 t$$
$$\psi_1^{(0)}(t) = t^2 (3-2t)$$

and

$$\psi_1^{(1)}(t) = t^2 (t-1).$$

The subscript kl denotes the value at the corner $p = k, q = l$.

It is an easy matter to see how the result (3.19) can be modified to give the required Hermite bicubic interpolating function over any rectangular element of the original rectangular type region. The approximating function over the complete region is then obtained in a manner similar to that used to obtain (1.13). This time there are *four* basis functions corresponding to each node of the rectangular array. For an internal node, i.e. a node not on the boundary of the rectangular type region, each basis function has a support of four rectangular elements. For a node on the boundary, but not at a corner, the support is two rectangular elements, and for the corner nodes one rectangular element. Over the complete rectangular region the piecewise bicubic approximating function gives $C^{1,1}$ continuity.[†]

A simplified version of (3.19) which is often used in practice involves only the values of H_3, $\partial H_3/\partial p$, and $\partial H_3/\partial q$ at each of the four corners. This, of course, is not unique, and an example of such a scheme which maintains quadratic precision (approximation exact for the functions $1, p, q, p^2, pq$, and q^2) but has now lost $C^{1,1}$ continuity is

$$H_3(p,q) = \sum_{j=1}^{4} \left[(H_3)_j \varphi_j(p,q) + \left[\frac{\partial H_3}{\partial p}\right]_j \psi_j(p,q) + \left[\frac{\partial H_3}{\partial q}\right]_j \chi_j(p,q) \right] \qquad (3.20)$$

[†]$C^{j,k}$ continuity of $u(x,y)$ means continuity of all the derivatives $D^{(L,M)} u (0 \leqslant L \leqslant j, 0 \leqslant M \leqslant k)$.

where
$$\varphi_1 = pq(-1 + 3p + 3q - 2p^2 - 2q^2)$$
$$\varphi_2 = q(p + 3q - 3p^2 - 2q^2 - 3pq + 2p^3 + 2pq^2)$$
$$\varphi_3 = 1 - 3p^2 - pq - 3q^2 + 2p^3 + 3p^2q + 3pq^2 + 2q^3 - 2p^3q - 2pq^3$$
$$\varphi_4 = p(3p + q - 2p^2 - 3pq - 3q^2 + 2p^2q + 2q^3)$$
$$\psi_1 = p^2q(p - 1)$$
$$\psi_2 = pq(p - 1)^2$$
$$\psi_3 = p(1 - q)(1 - p)^2$$
$$\psi_4 = p^2(p - 1)(1 - q)$$
$$\chi_1 = pq^2(q - 1)$$
$$\chi_2 = q^2(1 - p)(q - 1)$$
$$\chi_3 = (1 - p)q(q - 1)^2$$
$$\chi_4 = pq(q - 1)^2.$$

This can be constructed by the bilinear blending function technique of the previous section.

An interesting rectangular element has been described by Powell (1973). The rectangle is divided up into four triangles by the diagonals and a full quadratic expression in x and y assumed in each triangle. The coefficients in the quadratics can be chosen to permit C^1 continuity over a rectangular grid.

In concluding this short section on the rectangular element it is worth pointing out that whereas the size of the support of a spline increases with the order, the support of the Hermite function remains constant at four elements, irrespective of the order of the Hermite function, but the spline of course has continuity $C^{2(v-1),2(v-1)}$ as against $C^{v-1,v-1}$ for the Hermite function with polynomials of degree $2v - 1$.

Exercise 3.9 Using (3.19) show that the basis functions for $D^{(L,M)}U(0 \leqslant L, M \leqslant 1)$ at the node $(0,0)$ of a *unit mesh* are given by the *tensor product form*
$$\varphi^{(L,M)}(x, y) = \varphi^{(L)}(x)\varphi^{(M)}(y)$$
where
$$\varphi^{(0)}(t) = \begin{cases} (1 - t)^2(1 + 2t) & (0 \leqslant t \leqslant 1) \\ (1 + t)^2(1 - 2t) & (-1 \leqslant t \leqslant 0) \end{cases}$$
and
$$\varphi^{(1)}(t) = \begin{cases} (1 - t)^2t & (0 \leqslant t \leqslant 1) \\ (1 + t)^2t & (-1 \leqslant t \leqslant 0) \end{cases}$$

3.2.3 The quadrilateral

It might be thought that quadrilaterals are better mesh units than triangles because the overall grid is simplified. For example, a triangular network can always be simplified by combining the triangles in pairs to form quadrilaterals. Unfortunately, however, it is impossible to find a polynomial is x and y which reduces to an arbitrary linear form along the four sides of a general quadrilateral, and so it is not obvious how one can construct a piecewise function in x and y which has C^0 continuity over a quadrilateral network.

Lemma 3.1 *Let* $\mathscr{P}_1, \mathscr{P}_2, \mathscr{P}_3,$ *and* $\mathscr{P}_4$ *be points in* three-dimensional space *such that* $\mathscr{P}_j = (x_j, y_j, z_j)$ $(j = 1, 2, 3, 4)$. *Then the plane passing through the three points* $\mathscr{P}_j, \mathscr{P}_k,$ *and* $\mathscr{P}_l (j \neq k \neq l)$ *is*

$$\Pi_{jkl} = 0$$

where

$$\Pi_{jkl} = -zC_{jkl} + z_j D_{kl} - z_k D_{jl} + z_l D_{jk}$$

with $C_{jkl}, D_{kl},$ *etc. defined as in Section 3.1. In addition the surface*

$$\alpha \Pi_{klm} \Pi_{jkm} - \beta \Pi_{jlm} \Pi_{jkl} = 0 \tag{3.21}$$

where (j, k, l, m) *is any cyclic permutation of* $(1, 2, 3, 4)$, *passes through the four points* $\mathscr{P}_1, \mathscr{P}_2, \mathscr{P}_3$ *and* $\mathscr{P}_4$, *and contains the lines* $\mathscr{P}_1 \mathscr{P}_2, \mathscr{P}_2 \mathscr{P}_3, \mathscr{P}_3 \mathscr{P}_4$ *and* $\mathscr{P}_4 \mathscr{P}_1$ *for any values of* α *and* β.

From this lemma, it is clear that surfaces such as (3.21) that pass through points $\mathscr{P}_j = (x_j, y_j, f_j)$ $(j = 1, 2, 3, 4)$, can be used to define functions $f(x, y)$ in the quadrilateral $P_1 P_2 P_3 P_4$, where $P_j = (x_j, y_j)$, such that (i) $f(x_j, y_j) = f_j$ and (ii) $f(x, y)$ varies linearly along the sides $P_1 P_2, P_2 P_3, P_3 P_4$, and $P_4 P_1$ of the quadrilateral.

Thus it is possible to take $\mathscr{P}_j = (x_j, y_j, 1)$ and $\mathscr{P}_k = (x_k, y_k, 0)$ $(k \neq j)$ and so define a basis function $\varphi_j(x, y)$ such that

$$\varphi_j(x_k, y_k) = \begin{cases} 1 & (j = k) \\ 0 & (j \neq k) \end{cases} \quad (1 \leqslant j, k \leqslant 4). \tag{3.22}$$

Exercise 3.10 Let P_5 be the intersection of $P_1 P_2$ and $P_3 P_4$ and let P_6 be the intersection of $P_2 P_3$ and $P_4 P_1$ (Figure 3.5). Verify that the line $P_5 P_6 (D_{56} = 0)$ is such that

(i) $$\frac{D_{23}}{C_{123}} + \frac{D_{34}}{C_{134}} - \frac{D_{24}}{C_{124}} = \frac{D_{56}}{C_{156}} \tag{3.23a}$$

and that there exists some constant $\lambda \neq 0$ for which

(ii) $C_{134} C_{234} D_{12} + C_{123} C_{124} D_{34}$

$$= C_{123} C_{234} D_{14} + C_{134} C_{124} D_{23} = \lambda D_{56}. \tag{3.23b}$$

Hence show that if $\alpha C_{klm} C_{jkm} = \beta C_{jlm} C_{jkl}$ then the basis functions defined in (3.22) can be written as

$$\frac{D_{kl}}{C_{jkl}} \frac{D_{lm}}{C_{jlm}} \left(\frac{D_{56}}{C_{j56}} \right)^{-1}$$

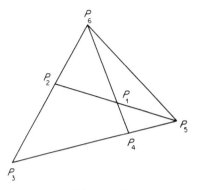

Figure 3.5

where $P_k P_l$ and $P_l P_m$ are the sides of the quadrilateral not containing the corner P_j, and where (j, k, l, m) is some permutation of $(1, 2, 3, 4)$. (*Hint* Use the formulae

(i) $C_{jkl} - C_{klm} + C_{jlm} - C_{jkm} = 0$ $\qquad\qquad\qquad$ (3.24a)

and

(ii) $C_{jkl} - D_{kl} + D_{jl} - D_{jk} = 0$ $\qquad\qquad\qquad$ (3.24b)

where (j, k, l, m) is any permutation of $(1, 2, 3, 4)$.)

3.3 Curved Boundaries

So far basis functions have been constructed in the main for networks with straight sides. In real problems in two and three dimensions, however, boundaries and interfaces are often curved. It is the purpose of this section to derive basis functions for networks composed of elements with curved sides (two dimensions) or curved surfaces (three dimensions).

3.3.1 Direct methods: triangle with one curved side

Initially we consider the triangular element with two straight sides and one curved side. This element together with triangles with straight sides can deal adequately with most plane problems involving curved boundaries and interfaces.

The triangle $P_1 P_2 P_3$ is considered in the (x, y) plane with $l(x, y) = 0$ and $m(x, y) = 0$ the equations of the straight sides $P_2 P_3$ and $P_3 P_1$ respectively. The equation of the curved side passing through P_1 and P_2 is $F(x, y) = 0$. We take $l(x, y)$, $m(x, y)$, and $F(x, y)$ to be normalized so that

$$l(x_1, y_1) = m(x_2, y_2) = F(x_3, y_3) = 1.$$

The transformation from the (x, y)-plane to the (l, m)-plane is given by

$$l = \frac{1}{C_{123}}(\tau_{23} + \eta_{23}x - \xi_{23}y) \quad m = \frac{1}{C_{123}}(\tau_{31} + \eta_{31}x - \zeta_{31}y) \quad (3.25\text{a})$$

or

$$x = x_3 - \zeta_{31}l + \zeta_{23}m \quad y = y_3 - \eta_{31}l + \eta_{23}m. \quad (3.25\text{b})$$

In the (l, m)-plane, the triangle becomes $P'_1 P'_2 P'_3$ where $P'_1 = (1, 0)$, $P'_2 = (0, 1)$, and $P'_3 = (0, 0)$, and the curved side $P'_1 P'_2$ is given by

$$F(x(l, m), y(l, m)) \equiv f(l, m) = 0.$$

We shall now described one type of Lagrangian approximation defined on the curved triangle $P'_1 P'_2 P'_3$ which satisfies the following conditions:

(i) Linear polynomials are interpolated exactly, that is,

$$\sum_i \varphi_i = 1; \quad \sum_i l_i \varphi_i = l; \quad \sum_i m_i \varphi_i = m, \quad (3.26)$$

where $\varphi_i(l, m)$ is a basis function associated with the node i. Since the transformation from (x, y) to (l, m) is linear (3.26) implies that linear polynomials in (x, y) on $P_1 P_2 P_3$ are also interpolated exactly, i.e. linear precision is maintained in (x, y)-space.

(ii) The basis function $\varphi_3(l, m)$ corresponding to P'_3 is identically zero on the curved side $P'_2 P'_1$.

(iii) The resulting piecewise smooth function, defined on a network of triangles, each triangle with at most one curved side, is C^0 continuous.

It is not difficult to see that in order to satisfy (i) and (ii) it is necessary to have at least four nodes in the triangle $P'_1 P'_2 P'_3$. In the simplest case these are taken as the three vertices together with an additional point $P'_4 = (l_4, m_4)$ on the curved side.

We first construct a basis function φ_3 that satisfies (ii) and then use (3.26) to construct φ_1, φ_2 and φ_4. We employ geometrical considerations similar to those adopted to obtain basis functions for the quadrilateral and so we consider the family of surfaces $z(l, m) = 0$ which intersect the (l, m) plane in the curve $f(l, m) = 0$ and are given by the equation

$$z(\alpha z + \beta l + \gamma m + \delta) + f(l, m) = 0. \quad (3.27)$$

If we impose conditions

$$(1) \ z = 1 \qquad (l = m = 0)$$
$$(2) \ z = 1 - l \quad (m = 0)$$
$$(3) \ z = 1 - m \quad (l = 0)$$

then it is possible to specify β, γ and δ in terms of an arbitrary constant α, such that (3.27) becomes

$$\alpha z^2 + \left[\alpha(l + m - 1) + 1 - \frac{f(l,0)}{1-l} - \frac{f(0,m)}{1-m} \right] z + f(l,m) = 0. \qquad (3.28)$$

We now identify z with φ_3 and the remaining basis functions are determined from (3.26) as

$$\varphi_1 = \frac{(1 - m_4)l + l_4 m - l_4}{1 - l_4 - m_4} + \frac{l_4}{1 - l_4 - m_4} \varphi_3$$

$$\varphi_2 = \frac{m_4 l + (1 - l_4)m - m_4}{1 - l_4 - m_4} + \frac{m_4}{1 - l_4 - m_4} \varphi_3 \qquad (3.29)$$

and

$$\varphi_4 = \frac{1 - l - m}{1 - l_4 - m_4} - \frac{1}{1 - l_4 - m_4} \varphi_3.$$

Exercise 3.11 Verify that if $\alpha = 0$ in (3.28) the basis function φ_3 becomes

$$\varphi_3 = \frac{f(l,m)}{f(0,m)/(1-m) + f(l,0)/(1-l) - 1} \qquad (3.30)$$

and further that if the curved side is part of the conic

$$f(l,m) = al^2 + blm + cm^2 - (1+a)l - (1+c)m + 1 = 0$$

then

$$\varphi_3 = \frac{al^2 + blm + cm^2 - (1+a)l - (1+c)m + 1}{1 - al - cm} \qquad (3.31)$$

and we recover the rational basis functions of Wachspress (1971, 1973a, b and 1975).

Exercise 3.12 If the curved side is part of a hyperbola

$$f(l,m) = blm - l - m + 1 = 0$$

show that the basis functions $\varphi_i (i = 1, 2, 3, 4)$ are polynomials if $\alpha = 0$.

   Note Piecewise hyperbolic arcs can be used to approximate a curved interface or boundary and still permit polynomial basis functions (McLeod and Mitchell, 1972, 1975).

Exercise 3.13 If Lagrange quadratics are used in a triangular network on a curved region, show that the approximant in a triangle adjacent to the boundary is

$$U(l,m) = l\left(1 - \frac{m}{m_4} - 2H \right) U_1 + m\left(1 - \frac{l}{l_4} - 2H \right) U_2$$

$$+ (1 - 2l - 2m)HU_3 + \frac{lm}{l_4 m_4} U_4 + 4mHU_5 + 4lHU_6$$

where the curved boundary segment has been replaced by a hyperbolic segment with equation

$$H(l, m) \equiv 1 - l - m + \frac{l_4 + m_4 - 1}{l_4 m_4} lm = 0$$

and the numbering of the nodes is as in Figure 3.1. Verify that the linear precision conditions (3.26) are satisfied.

3.3.2 Isoparametric transformations

In this chapter we have already given several examples of the general Lagrange interpolant

$$U(\mathbf{s}) = \sum_j U_j \varphi_j(\mathbf{s}) \tag{3.32}$$

where $\mathbf{s}$ is the coordinate vector (p, q) (or (p, q, r)) in 2 (or 3) dimensions in a *reference* (*standard*) element which is either triangular (or tetrahedral) or rectangular (or hexahedral), and $\varphi_j(\mathbf{s})$ is a basis function, usually polynomial, which is unity at the jth node of the element and zero at all other nodes. The point transformation (Irons, 1966; Zienkiewicz, 1977)

$$t = \sum_j t_j \varphi_j(\mathbf{s}), \quad t = x, y, (z) \tag{3.33}$$

which has the same form as (3.32) defines an *isoparametric* transformation, and it follows immediately from (3.33) that *linear precision* is maintained in each element in $x, y, (z)$ space. This can be expressed in the form

$$\sum_j \varphi_j = 1; \quad \sum_j x_j \varphi_j = x; \quad \sum_j y_j \varphi_j = y; \quad \left(\sum_j z_j \varphi_j = z \right), \tag{3.34}$$

where $\varphi_j(x, y, (z))$ is a basis function associated with the node j.

Associated with the transformation (3.33) is the Jacobian

$$J = \frac{\partial(x, y, (z))}{\partial(p, q, (r))} \tag{3.35}$$

which must satisfy the condition

$$J > 0$$

inside the standard element in $p, q, (r)$ space, and which also figures prominently in the integrals central to the solution of any problem by finite elements.

We now give several examples of isoparametric transformations based on Lagrange interpolants in the (p, q) plane.

3.3.3 The unit square reference element

(1) *Bilinear approximation*   The standard transformation is

$$t = pqt_1 + (1 - p)qt_2 + (1 - p)(1 - q)t_3 + p(1 - q)t_4 \quad (t = U, x, y) \tag{3.36}$$

which takes the corner points $P_j \equiv (x_j, y_j)(j = 1, 2, 3, 4)$ of a quadrilateral into the four corners $(1, 1)$, $(0, 1)$, $(0, 0)$ and $(1, 0)$ of the unit square in (p, q) space.

If the bilinear polynomials used in the transformation (3.36) are replaced by polynomials of higher degree, it is possible to introduce additional points in the transformation and at the same time extend the isoparametric approximations to curvilinear quadrilaterals.

(2) *Biquadratic approximation* The approximation and associated isoparametric transformations are given by

$$t = \sum_{j=1}^{9} \varphi_j^{(2)}(p, q)t_j \quad (t = U, x, y) \tag{3.37}$$

where

$$\varphi_1^{(2)} = p(2p - 1)q(2q - 1)$$

with $\varphi_2^{(2)}$, $\varphi_3^{(2)}$ and $\varphi_4^{(2)}$ defined similarly,

$$\varphi_5^{(2)} = 4p(1 - p)q(2q - 1)$$

with $\varphi_6^{(2)}$, $\varphi_7^{(2)}$ and $\varphi_8^{(2)}$ defined similarly, and

$$\varphi_9^{(2)} = 16p(1 - p)q(1 - q).$$

N.B. If the transformation $(t = x, y)$ is defined by (3.36) and the approximation $(t = U)$ by (3.37), we have an example of a *subparametric* transformation.

In Section 3.2.1 the internal node P_9 is eliminated by using the linear relation

$$U_9 = \tfrac{1}{2} \sum_{j=5}^{8} U_j - \tfrac{1}{4} \sum_{j=1}^{4} U_j. \tag{3.38}$$

This yields a function that still recovers quadratics in p and q exactly but has no term in $p^2 q^2$ in the approximation which can be written as

$$U(p, q) = \sum_{j=1}^{8} \varphi_j^{(2)*}(p, q)U_j \tag{3.39}$$

where $\varphi_j^{(2)*}$ are given in (3.16b). The eight-mode isoparametric element is thus defined by

$$t = \sum_{j=1}^{8} \varphi_j^{(2)*}(p, q)t_j \quad (t = U, x, y).$$

(3) *Bicubic approximation* The full bicubic approximation involves four internal nodes. These can be eliminated to yield the approximation and associated isoparametric transformation

$$t = \sum_{j=1}^{12} \varphi_j^{(3)*}(p, q)t_j \quad (t = U, x, y) \tag{3.40}$$

where the $\varphi_j^{(3)*}$ are defined either by (3.18b) on (3.18c).

Exercise 3.14 Verify that the inverse transformation of (3.36) can be written as

$$(C_{234} + C_{134})p^2 + (D_{34} + D_{12} - C_{123} - C_{234})p + D_{23} = 0 \tag{3.41a}$$

and

$$(C_{234} + C_{123})q^2 + (D_{23} + D_{41} - C_{134} - C_{234})q + D_{34} = 0. \tag{3.41b}$$

Further, by using (3.24a) and (3.24b) show that the function p defined by (3.41a) is equivalent to a surface of the form (3.21) through the points $\mathscr{P}_j = (x_j, y_j, f_j)$ with $f_2 = f_3 = 0$ and $f_1 = f_4 = 1$, if $\alpha = \beta$, and that q defined by (3.37b) is equivalent to a surface with $f_3 = f_4 = 0$, $f_1 = f_2 = 1$ if $\alpha = \beta$.

Exercise 3.15 The particular choice $\alpha = \beta$ is not the only one possible if a transformation from the quadrilateral $P_1 P_2 P_3 P_4$ to the unit square is required. Show that if $\mathscr{P}_j (j = 1, 2, 3, 4)$ are chosen as in Exercise 3.14 and $\alpha C_{klm} C_{jkm} = \beta C_{jlm} C_{jkl}$, then the new coordinates p and q can be defined by

$$p = \frac{D_{23}}{C_{123}} \frac{D_{34}}{C_{134}} \left(\frac{D_{56}}{C_{156}} \right)^{-1} + \frac{D_{23}}{C_{234}} \frac{D_{12}}{C_{124}} \left(\frac{D_{56}}{C_{456}} \right)^{-1}$$

and

$$q = \frac{D_{23}}{C_{123}} \frac{D_{34}}{C_{134}} \left(\frac{D_{56}}{C_{156}} \right)^{-1} - \frac{D_{34}}{C_{234}} \frac{D_{14}}{C_{124}} \left(\frac{D_{56}}{C_{256}} \right)^{-1}.$$

Exercise 3.16 Show that the Jacobian J of the transformation defined by (3.36) can be written as

$$J = (1 - p)C_{123} + (1 - q)C_{134} + (p + q - 1)C_{124}$$

and verify that $J > 0$ for $0 \leqslant p, q \leqslant 1$ if P_1, P_2, P_3 and P_4 are numbered anticlockwise.

3.3.4 The hexahedron

The element having six quadrilateral faces is in general a better element in three dimensions than the tetrahedron. Local isoparametric coordinates (p, q, r) can be introduced as for quadrilateral elements in two dimensions. The transformations are given by

$$t = \sum_{j=1}^{8} \varphi_j^{(1)}(p, q, r) t_j \quad (t = x, y, z) \tag{3.42}$$

where

$$\varphi_1^{(1)} = pqr$$
$$\varphi_2^{(1)} = (1 - p)qr$$

etc., where the vertices are numbered as in Figure 3.6. An arbitrary hexahedron is thus transformed into the unit cube in (p, q, r) space. The isoparametric approximation is then defined by

$$U(p, q, r) = \sum_{j=1}^{8} \varphi_1^{(1)}(p, q, r) U_j. \tag{3.43}$$

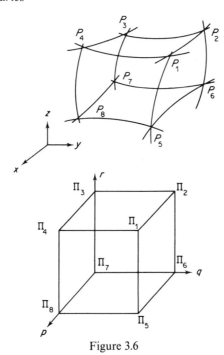

Figure 3.6

Triquadratic and tricubic approximations can be generated if additional points on the sides and the faces are introduced, together with a number of interior points. The face points and interior points can be eliminated in a manner analogous to that used in two dimensions to eliminate the interior points of a quadrilateral.

Exercise 3.16 Compute the basis functions $\varphi_j^{(2)}(p,q,r)\,(j=1,\ldots,27)$ for the triquadratic isoparametric approximation defined for a hexahedron. Then verify that the centre face points and the centroid can be eliminated to yield an approximation of the form

$$U(p,q,r) = \sum_{j=1}^{20} \varphi_j^{(2)^*}(p,q,r)U_j \tag{3.44}$$

which interpolates quadratics exactly, has no terms in $p^2q^2, q^2r^2, r^2p^2, p^2q^2r, p^2qr^2, pq^2r^2, p^2q^2r^2$, and for which

$$\varphi_1^{(2)^*} = pqr(2p + 2q + 2r - 5)$$

and similarly for $\varphi_j^{(2)^*}(j=2,\ldots,8)$,

$$\varphi_9^{(2)^*} = 4pqr(1-p)$$

with $\varphi_j^{(2)^*}(j=10,\ldots,20)$ similarly.

Exercise 3.17 Verify that tricubic approximation in a hexahedron requires 64 nodes, one at each corner with a further two on each edge, four on each face, and eight in the interior of the hexahedron. Then verify that the face points and interior points can be eliminated to yield an approximation that involves terms in $p^j q^k r^l (j + k + l \leqslant 4)$ together with the three quintic terms $p^3 qr, pq^3 r$ and pqr^3, such that

$$U(p, q, r) = \sum_{j=1}^{32} \varphi_j^{(3)^*}(p, q, r) U_j \tag{3.45}$$

where

$$\varphi_1^{(3)^*} = \tfrac{9}{2} pqr[p^2 + q^2 + r^2 - (p + q + r) + \tfrac{2}{9}]$$

with $\varphi_j^{(3)^*} (j = 2, \ldots, 8)$ similarly, and

$$\varphi_9^{(3)^*} = \tfrac{9}{2} pqr(1 - p)(3p - 1)$$

with $\varphi_j^{(3)^*} (j = 10, \ldots, 32)$ similarly.

3.3.5 The quadratic triangle

The approximation and associated isoparametric transformation of Figure 3.7 are given by

$$t = \sum_{j=1}^{6} \varphi_j^{(2)}(p, q) t_j \quad (t = U, x, y) \tag{3.46}$$

where

$$\varphi_1^{(2)} = p(2p - 1) \quad \varphi_2^{(2)}, \; \varphi_3^{(2)} \text{ similarly}$$

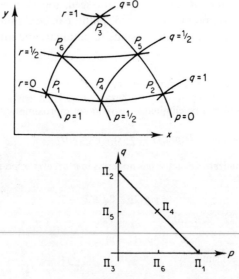

Figure 3.7

and

$$\varphi_4^{(2)} = 4pq \qquad \varphi_5^{(2)}, \ \varphi_6^{(2)} \text{ similarly.}$$

If we consider the special case of a triangle with two straight sides and one curved side such that $t_5 = \frac{1}{2}(t_2 + t_3), t_6 = \frac{1}{2}(t_3 + t_1) (t = x, y)$, the transformation (3.46) can be used to give the simplified formulae

$$l = \alpha pq + p \tag{3.46a}$$

and

$$m = \beta pq + q \tag{3.46b}$$

where $\alpha = 2(2l_4 - 1)$, $\beta = 2(2m_4 - 1)$ and l, m are as given by (3.25). The inverse transformation is given by

$$\beta p^2 + (\alpha m - \beta l + 1)p - l = 0 \tag{3.47a}$$

and

$$\alpha q^2 + (\beta l - \alpha m + 1)q - m = 0. \tag{3.47b}$$

The quadratic isoparametric transformation (3.46) is now used to obtain a piecewise parabolic C^1 approximation to a curved boundary. Each parabolic arc has four free parameters and these are used up by specifying the position and slope at the ends of the arc. Thus the approximation is local in nature and so preserves the convexity/concavity of the original curve. Now consider a parabolic arc between the nodes (x_i, y_i) and (x_{i+1}, y_{i+1}) on the given curved boundary to be given by

$$x = as^2 + bs + c$$
$$y = ds^2 + es + f \tag{3.48}$$

where a, b, c, d, e, and f are constant coefficients and $s \in [0, 1]$. The given curve is taken to be C^3 and between the nodes its equation is

$$x = A(t) \quad y = B(t)$$

where $t \in [t_i, t_{i+1}]$. From (3.48), Hermite interpolation of position and slope at the endpoints of the arc leads to

$$x_i(s) = -\alpha_i s^2 + (\alpha_i + x_{i+1} - x_i)s + x_i$$
$$y_i(s) = -\beta_i s^2 + (\beta_i + y_{i+1} - y_i)s + y_i \tag{3.49}$$

with

$$\alpha_i = \frac{2(y_{i+1} - y_i)f_i f_{i+1} + (x_i - x_{i+1})(f_i + f_{i+1})}{f_i - f_{i+1}}$$

and

$$\beta_i = \frac{2(x_i - x_{i+1})f_i f_{i+1} - (y_i - y_{i+1})(f_i + f_{i+1})}{f_i - f_{i+1}}$$

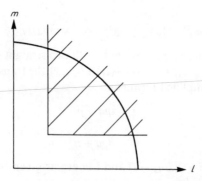

Figure 3.8

where f_i and f_{i+1} are the slopes of the parabolic arc at the nodes (x_i, y_i) and (x_{i+1}, y_{i+1}) respectively, and for a parabola $f_i \neq f_{i+1}$. The parabola given by (3.49) can also be obtained from the isoparametric transformation (3.46) where the intermediate point (X, Y) on the curved side has been chosen to satisfy

$$X = \tfrac{1}{4}\alpha_i + \tfrac{1}{2}(x_i + x_{i+1}), \quad Y = \tfrac{1}{4}\beta_i + \tfrac{1}{2}(y_i + y_{i+1}).$$

Further details of matching curved boundaries with C^1 approximations can be found in McLeod and Mitchell (1979).

One of the difficulties in dealing with curved elements using isoparametric coordinates arises from the vanishing of the Jacobian $J(= 1 + \beta p + \alpha q)$ of the transformation defined by (3.46a) and (3.46b). This Jacobian is positive for all p, q such that $0 \leq p, q$ and $p + q \leq 1$ provided that the point (l_4, m_4) lies in the region $l, m > \tfrac{1}{4}$ as shown in Figure 3.8. For other positions of the point (l_4, m_4) in the positive quadrant of the (l, m) plane, including the lines $l = \tfrac{1}{4}$, $m = \tfrac{1}{4}$, the Jacobian either vanishes or is negative for certain values of (p, q) (Jordan, 1970) and so isoparametric coordinates cannot in general be used to deal with curved elements in these 'forbidden' cases (exceptions to this rule are given in Section 7.6). This is because results calculated in terms of the isoparametric coordinates (p, q) cannot be transferred back to the (l, m) plane because of the vanishing of the Jacobian of the transformation somewhere in the element.

Instead of using isoparametric coordinates, we can deal directly in terms of l and m by using (3.28) and (3.29). As in Exercise 3.18, if the curved side is given by (3.48) and $\alpha = (2R - 1)/2R$, then (3.28) will have real roots provided

$$F(l, m; R) = \left(l + m + \frac{1}{2(2R - 1)} \right)^2 - 4lm \geq 0. \tag{3.50}$$

It is easy to see that for a fixed value of R the function $F(l, m; R)$ has no maximum or minimum inside the element, or indeed anywhere in the (l, m)-plane. Consequently the smallest value of $F(l, m; R)$ will occur on the boundary of the

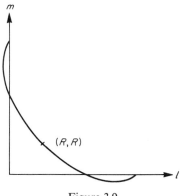

Figure 3.9

element for all values of R. In fact, using (3.48), it follows that

$$F(l, m; R) = \left(\frac{4R - 1}{4R - 2}\right)^2 \geqslant 0$$

for all R on the curved side, and of course F is always positive from (3.50) on $l = 0$ and $m = 0$. The condition (3.50) is thus satisfied for all values of R and for all points (l, m) in the element.

When isoparametric coordinates are used in this example, values of R in the range $0 < R < \frac{1}{4}$ are forbidden. This does not seem to be the case when (3.29) is used. However, by simple geometrical considerations, it can be shown that for $0 < R < \frac{1}{4}$, the curved boundary intersects the l and m axes at points between the origin and the unit points (Figure 3.9).

Exercise 3.18 For $0 < R < \frac{1}{4}$, find the points on the l and m axes between the origin and the unit points where the curved side intersects the axes, and show that as $R \to \frac{1}{4}$ these points tend towards the unit points on the respective axes.

Exercise 3.19 Show that the quadratic equation (3.28) has real roots for all values of α when (l, m) is a point on the perimeter of the triangle with two straight sides and one curved side.

Exercise 3.20 Show that if $l_4 = m_4 = R$ then it follows from (3.47a) and (3.47b) that $r(= 1 - p - q)$ is given by

$$r^2 - \frac{4R - 1}{2R - 1}r + \left[\frac{2R - l - m}{2R - 1} - (l - m)^2\right] = 0 \qquad (3.51a)$$

and hence that the curved side $f(l, m) = 0$ is replaced by the curve $r = 0$ given by

$$1 - \frac{l + m}{2R} - \frac{2R - 1}{2R}(l - m)^2 = 0 \qquad (3.51b)$$

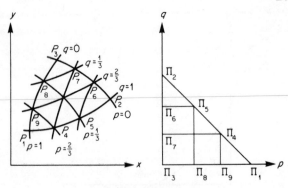

Figure 3.10

which is a parabola. Further show that if $\alpha = (2R - 1)/2R$ and $f(l, m)$ is given by (3.51b) then (3.28) becomes (3.51a) with r replacing z. Explain this link between isoparametric approximation and direct methods of dealing with curved boundaries.

Exercise 3.21 Show that in general the curved side $r = 0$ is given by

$$(\beta l - \alpha m)^2 + (\alpha + \beta + \alpha\beta - \beta^2)l + (\alpha + \beta + \alpha\beta - \alpha^2)m - (\alpha + \beta + \alpha\beta) = 0$$

where α and β are as given in (3.47a) and (3.47b).

3.3.6 Cubic approximation

The approximation and associated isoparametric transformation of Figure 3.10 are given by

$$t = \sum_{j=1}^{9} \varphi_j^{(3)}(p, q)t_j \quad (t = U, x, y) \tag{3.52}$$

where the basis functions $\varphi_j^{(3)}(p, q)$ are the functions $\bar{p}_j^{(s)}$ given by (3.6). The approximation has quadratic precision in p and q and linear precision in x and y, the latter being guaranteed by the isoparametric transformation.

We now consider a typical boundary element with one curved side and two straight sides, the nodes on the straight sides being placed at the points of trisection and those on the curved side placed arbitrarily at the points (x_4, y_4) and (x_5, y_5). On the curved side we have

$$p + q = 1$$

and so from (3.52) the parametric formulae for the curved side become

$$x(p, q) = x_2 + (x_1 - \tfrac{11}{2}x_2 - \tfrac{9}{2}x_4 + 9x_5)p - \tfrac{9}{2}(x_1 - 2x_2 - 4x_4 + 5x_5)p^2$$
$$+ \tfrac{9}{2}(x_1 - x_2 - 3x_4 + 3x_5)p^3 \tag{3.53}$$

and a similar formula for $y(p, q)$, where $0 \leqslant p \leqslant 1$.

We look upon the points (x_1, y_1) and (x_2, y_2) as the fixed points (x_i, y_i) and (x_{i+1}, y_{i+1}) on the original curve, and the points (x_4, y_4) and (x_5, y_5) as movable points on the element boundary which we attempt to place to suit specific requirements. The *cubic curve* (3.52) can sustain cusps, loops, etc. and so may, in general, be a dangerous replacement for the original curved arc. Suitable placements of the points (x_4, y_4) and (x_5, y_5) are discussed in Woodford *et al.* (1978) and Mitchell (1979); in particular if

$$l_5 = l_4 - \tfrac{1}{3}$$

and

$$m_5 = m_4 + \tfrac{1}{3}$$

then the *cubic curve degenerates into a unique parabola* through the four points $(1, 0)$, (l_4, m_4), (l_5, m_5), and $(0, 1)$ in the (l, m)-plane. The connecting formulae between the (x, y) and (l, m)-planes are given by (3.25).

The Jacobian of the isoparametric transformation can also cause problems in the cubic case; a discussion of this can be found in Woodford *et al.* (1978), Mitchell (1979), and Brown and Wait (1982).

The Jacobian The main advantage of isoparametric transformations for dealing with curved boundaries is the ease with which a basis is constructed. A substantial disadvantage, however, occurs in the integration which is carried out in a space other than that in which the problem is defined.

If for example the transformation is given by

$$x = x(p, q), \quad y = y(p, q) \tag{3.54}$$

and the shape functions $u(p, q)$ are polynomial functions of p and q, then the integrals requiring evaluation are a selection of

$$\iint P_1(p, q) J \, dp dq \qquad \text{element of mass matrix}$$

$$\iint P_2(p, q) \, dp dq \qquad \text{element of convection matrix}$$

$$\iint P_3(p, q) J^{-1} \, dp dq \qquad \text{element of stiffness matrix}$$

where P_1, P_2 and P_3 are polynomials. Thus as J is a polynomial, rational integrands can arise even although a polynomial basis is used.

It is also worth observing that the Jacobian of the isoparametric transformation is given by

$$J = \frac{\partial(x, y, (z))}{\partial(l, m, (n))} \cdot \frac{\partial(l, m, (n))}{\partial(p, q, (r))} \tag{3.55}$$

and since the transformation from $(x, y, (z))$ to $(l, m, (n))$ is *linear* the value of the first Jacobian in (3.55) is constant, and so the important part of (3.55) is the second Jacobian. For example in the case of a triangle with one curved side,

$$\frac{\partial(x, y)}{\partial(l, m)} = C_{123} \quad \text{(from (3.25b))}.$$

Exercise 3.22 Starting with the transformation given by (3.54) obtain the result

$$\begin{vmatrix} p_x & p_y \\ q_x & q_y \end{vmatrix} = \frac{1}{J} \begin{vmatrix} y_q & -x_q \\ -y_p & x_p \end{vmatrix}.$$

Hence show for the function $\varphi(x, y)$, with x and y given by (3.54), that

$$J\frac{\partial \varphi}{\partial x} = y_q \frac{\partial \varphi}{\partial p} - y_p \frac{\partial \varphi}{\partial q}$$

and similarly for $\partial \varphi / \partial y$.

3.3.7 High-order bases

So far in the direct method and the isoparametric transformation methods we have achieved only *linear precision* in dealing with curved boundaries. In order to achieve greater 'accuracy' in the finite element method in general, the choice is either to reduce the size of the element $(h \to 0)$ or to keep the element size fixed and to increase the order of the polynomial basis. The remainder of this section is devoted to the construction of high-order bases.

It should be said at the outset that techniques for constructing bases of any desired degree of precision in the original plane already exist for a wide class of two- and three-dimensional curved elements (Wachspress, 1975). Unfortunately the resulting basis functions are rational and so far finite element users have avoided such a procedure. We shall describe an alternative high-order method due to McLeod (1978, 1979) which uses the same type of nonlinear transformation as the isoparametric method but employs a different basis construction. The latter requires a knowledge of the number of extra nodes and associated basis functions required on the curved side in order to increase the precision in the element to the desired level. The answer lies in the formula

$$N = \tfrac{1}{2}[(n + 1)(n + 2) - (n - m + 1)(n - m + 2)] \tag{3.56}$$

where N is the number of nodes required on an algebraic curve of degree m to span polynomials of degree n.

We shall now give an example of McLeod's high-order transformation method of increasing precision in a triangular element bounded by two straight sides and a conic arc. It will be assumed that a linear transformation has already taken place and that the element is in the (l, m)-plane and has *quadratic* precision. Each straight side has $m = 1$, $n = 2$ and so, from (3.56), $N = 3$. For the conic arc, $m =$

$n = 2$ and so $N = 5$. Hence two additional nodes are required on the conic arc, leading to an approximant in the (l, m)-plane

$$U(l, m) = \sum_{j=1}^{8} U_j \varphi_j(p, q) \tag{3.57}$$

where $l = l(p, q)$, $m = m(p, q)$ and the conditions for quadratic precision are

$$\sum_{j=1}^{8} \varphi_j = 1; \quad \sum_{j=1}^{8} l_j \varphi_j = l; \quad \sum_{j=1}^{8} m_j \varphi_j = m;$$

$$\sum_{j=1}^{8} l_j^2 \varphi_j = l^2; \quad \sum_{j=1}^{8} l_j m_j \varphi_j = lm; \quad \sum_{j=1}^{8} m_j^2 \varphi_j = m^2. \tag{3.58}$$

Although the high-order transformation method is not restricted to any particular transformation it is convenient for us to assume the isoparametric transformation given by (3.46a) and (3.46b) which maps a parabolic arc in the (l, m)-plane into a straight line in the (p, q)-plane. The extra nodes Π_7 and Π_8 in Figure 3.11(b), located at the points $(\frac{3}{4}, \frac{1}{4})$ and $(\frac{1}{4}, \frac{3}{4})$ respectively, are transformed by (3.46a) and (3.46b) into nodes P_7 and P_8 on the parabolic arc in Figure 3.11(a). The high-order basis is then obtained by first constructing

$$\varphi_7 = \frac{16}{3} pq \frac{D_{84}}{C_{784}},$$

$$\varphi_8 = \frac{16}{3} pq \frac{D_{74}}{C_{874}} \tag{3.59}$$

where, in the notation of Section 3.1, D_{jk}/C_{ijk} is a normalized linear form in (l, m)-coordinates. The remaining φ's, obtained from (3.58), can be rewritten in the form

$$\varphi_j(l, m; p, q) = Q_j(l, m) - Q_j(l_7, m_7)\varphi_7(l, m; p, q)$$
$$- Q_j(l_8, m_8)\varphi_8(l, m; p, q) \quad j = 1, 2, \dots, 6 \tag{3.60}$$

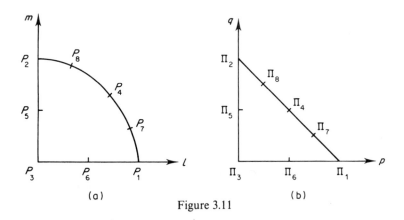

Figure 3.11

where $Q_j(l,m), j = 1, 2, \ldots, 6$, denotes the quadratic polynomial which is zero at the nodes $(l_i, m_i), i \neq j$, and has unit value at (l_j, m_j).

3.4 Nonconforming Elements

So far the overall approximation in a finite element method has required a certain amount of *inter-element continuity*. For an elliptic equation of order $2k(k = 1, 2)$, this requirement is either C^{k-1} for Ritz and Galerkin, or C^{2k-1} for least squares. When one considers that ninth-order polynomials are required to obtain C^1 continuity between tetrahedral elements, it is obvious that for $k = 2$ the application of elements with the required amount of continuity, i.e. conforming elements, may be a formidable task. From a computational point of view, it is thus desirable to use elements with less continuity than appears to be required, i.e. nonconforming elements.

Examples of nonconforming elements are, for *second*-order problems:

(1) The de Veubeke element. This is a triangle where the linear polynomial interpolates function values at the midpoints of the triangle sides.
(2) The Wilson element (Wilson *et al.*, 1971). On the square $0 \leqslant x, y \leqslant 1$, the six basis functions consist of $xy, x(1-y), y(1-x), (1-x)(1-y)$, together with the additional functions $4x(1-x)$ and $4y(1-y)$. The latter functions make it possible to produce any bivariate quadratic polynomial and so permit an improved representation within each element.

For *fourth*-order problems:

(3) The Morley triangle where a quadratic polynomial interpolates function values at the vertices and normal derivatives at the midpoints of the sides.
(4) The Adini element (Adini and Clough, 1961). This twelve degree of freedom element has $u, \partial u/\partial x, \partial u/\partial y$ at the corners of the rectangle as unknown parameters and the complete cubic together with $x^3 y$ and xy^3 as basis functions (see Section 3.2.2).

4

Methods of Approximation

4.1 Introduction

Finite element methods can be formulated in two distinct ways, one involving variational principles, the other not. *Ritz methods*,[†] based on the approximate solution of variational problems, appear most frequently in structural analysis and in the non finite element literature (Courant and Hilbert, 1953; Nečas, 1967). On the other hand variants of the *Galerkin methods*[††] form the basis of most nonstructural finite element calculations.

In either case we require an approximate solution in the form

$$U(\mathbf{x}) = \sum_i U_i \varphi_i(x) \tag{4.1}$$

where the nodal parameters U_i may include derivative values if a Hermite formulation is adopted. The basis functions φ_i are piecewise smooth (occasionally only piecewise continuous) functions, constructed element-by-element from the shape functions detailed in the previous chapter. If we denote by φ^e the restriction of φ to the individual element e then

$$\varphi_i = \sum_e \varphi_i^e. \tag{4.2}$$

In addition we assume that, in order to facilitate the integration, each element is mapped onto a reference element using a differentiable (isoparametric) transformation.

[†]Named after the Swiss mathematician Walter Ritz (1878–1909).

[††]After the Russian mathematician Boris G. Galerkin (1871–1945), although in many Russian texts it is referred to as the Bubnov–Galerkin method to acknowledge the independent contributions of Galerkin (in 1915) and I. G. Bubnov (in 1913).

Thus integrals of the form

$$\iint_R f(U)\,dx\,dy$$

are computed as

$$\sum_e \iint_e f(U)\frac{\partial(x,y)}{\partial(p,q)}\,dp\,dq \tag{4.3}$$

where $\partial(x,y)/\partial(p,q)$ represents the Jacobian

$$J = \det \begin{bmatrix} \dfrac{\partial x}{\partial p} & \dfrac{\partial x}{\partial q} \\[2mm] \dfrac{\partial y}{\partial p} & \dfrac{\partial y}{\partial q} \end{bmatrix} \tag{4.4}$$

of the transformation from the element in physical (x,y) space to the reference element in (p,q) space. Similarly, if the integrand involves derivatives these have to be transformed; thus $\partial u/\partial x$ is replaced by

$$\frac{\partial u^e}{\partial p}\frac{\partial p}{\partial x} + \frac{\partial u^e}{\partial q}\frac{\partial q}{\partial x} \tag{4.5}$$

and so on.

Note that if in each element isoparametric transformations of the form

$$\begin{bmatrix} u^e \\ x \\ y \end{bmatrix} = \sum_i \begin{bmatrix} u_i \\ x_i \\ y_i \end{bmatrix} \varphi_i^e(p,q) \tag{4.6}$$

are used, then

$$\frac{\partial x}{\partial p} = \sum_i x_i \frac{\partial \varphi_i}{\partial p}$$

etc. and the derivatives such as $\partial p/\partial x$ are computed as elements of J^{-1} (Exercise 3.22).

4.1.1 Ritz methods

Assume we require an approximate solution of the variational problem

$$\min_{u \in \mathcal{H}} I(u). \tag{4.7}$$

If the approximate solution is as in (4.1), then minimizing over the N-dimensional subspace spanned by the basis functions φ_i leads to the approximate problem

$$\frac{\partial}{\partial U_i} I\left(\sum_j U_j \varphi_j\right) = 0 \quad (i = 1,\ldots,N).$$

Thus we have one equation corresponding to each parameter *not otherwise specified by the boundary data*. If for example we are solving

$$\min\left\{\int\int_R \tfrac{1}{2}(u_x^2 + u_y^2)\,dxdy - \int\int_R uf\,dxdy\right\}$$

this technique leads to the approximate problem

$$\int\int_R \frac{1}{2}\frac{\partial}{\partial U_i}\left\{\left(\frac{\partial U}{\partial x}\right)^2 + \left(\frac{\partial U}{\partial y}\right)^2 - 2Uf\right\}dxdy = 0 \quad (i = 1, \dots, N). \tag{4.8}$$

Using (4.1), this can be rewritten as

$$\int\int_R \sum_j U_j\left\{\left(\frac{\partial \varphi_j}{\partial x}\right)\left(\frac{\partial \varphi_i}{\partial x}\right) + \left(\frac{\partial \varphi_j}{\partial y}\right)\left(\frac{\partial \varphi_i}{\partial y}\right)\right\}dxdy = \int\int_R \varphi_i f\,dxdy$$
$$(i = 1, \dots, N). \tag{4.9}$$

Using (4.3), etc. the left-hand side can be written as

$$\sum_e \sum_j U_j \int\int_e \left\{\left(\frac{\partial \varphi_j}{\partial x}\right)^e\left(\frac{\partial \varphi_i}{\partial x}\right)^e + \left(\frac{\partial \varphi_j}{\partial y}\right)^e\left(\frac{\partial \varphi_i}{\partial y}\right)^e\right\}\frac{\partial(x, y)}{\partial(p, q)}\,dpdq \quad (i = 1, \dots, N)$$
$$\tag{4.10}$$

where

$$\frac{\partial \varphi_j^e}{\partial x} \equiv \left(\frac{\partial \varphi_j}{\partial p}\right)^e\left(\frac{\partial p}{\partial x}\right) + \left(\frac{\partial \varphi_j}{\partial q}\right)^e\left(\frac{\partial q}{\partial x}\right) \tag{4.11}$$

etc. The right-hand side is also expressed in terms of summation $\sum_e \int\int_e$.

It was shown in Chapter 2 that the solution of the variational problem (4.7) is also the solution of the boundary value problem comprising the Euler–Lagrange equation and certain boundary conditions. These latter may be either forced boundary conditions imposed on $\mathscr{H}$ or they may be natural boundary conditions that are only satisfied at the solution (see Section 2.3). In the preceding example, assume that we take $\mathscr{H} = \mathscr{H}^{(1)}(R)$ and impose no additional restrictions on $\mathscr{H}$. It follows that the solution of (4.7) satisfies the natural boundary condition

$$\frac{\partial u}{\partial n} = 0$$

and the approximate solution (4.1) with the parameters U_i, specified by (4.8), will approximate this condition as well as approximating the Euler–Lagrange equation

$$-\frac{\partial^2 u}{\partial x^2} - \frac{\partial^2 u}{\partial y^2} = f \quad \text{in } R. \tag{4.12}$$

For the approximation solution to have a reasonable degree of accuracy it is vital that at least some of φ_i can take nonzero values on the boundary ∂R.

Exercise 4.1 Assuming that the boundary ∂R can be written as $\partial R = \partial R_1 + \partial R_2$ and the boundary conditions are

$$u = 0 \quad \text{on } \partial R_2$$

and

$$\frac{\partial u}{\partial n} = g \quad \text{on } \partial R_1$$

what is the functional $I(u)$ if the Euler–Lagrange equation remains (4.12)?

Computational considerations

The order of the summations in (4.10) is important from a computational point of view. All the integrals relating to an individual element are computed together. Then the coefficient of U_j in equation i of (4.9) is assembled element-by-element. For each k-node element e, we construct the $k \times k$ matrix G^e, with components (in the preceding example)

$$a_{ij}^e = \iint_e \left\{ \left(\frac{\partial \varphi_i}{\partial x}\right)^e \left(\frac{\partial \varphi_j}{\partial x}\right)^e + \left(\frac{\partial \varphi_i}{\partial y}\right)^e \left(\frac{\partial \varphi_j}{\partial y}\right)^e \right\} \mathrm{d}x\mathrm{d}y. \tag{4.13}$$

These include only those φ_i and φ_j that are nonzero in e. The components of the global matrix A are then assembled piecemeal as

$$a_{ij} = \sum_e a_{ij}^e. \tag{4.14}$$

This element-by-element assembly can be interwoven with the solution by Gaussian elimination to yield the so-called *frontal method* of solution. This and other methods of solution are discussed in Section 4.5.

4.2 Galerkin Methods

The existence of an Euler–Lagrange equation follows as a necessary condition for the solution of a variational problem. The converse is not true, i.e. the existence of a differential equation does not guarantee the existence of an equivalent variational principle. Thus it is necessary to have a definition of approximations of the form (4.1) that does not depend on the variational formulation.

4.2.1 Weak solutions

A function $u(\mathbf{x})$ that satisfies the linear differential equation

$$Au = f \quad \mathbf{x} \in R \tag{4.15}$$

is called a *classical solution*, to distinguish it from a *weak* solution that may satisfy only

$$(Au, v) = (f, v) \qquad (4.16)$$

or

$$(u, A^*v) = (f, v) \qquad (4.17)$$

for all admissible *test functions* v, where (,) is the $\mathscr{L}_2$-inner product

$$(u, v) = \int\!\!\int_R uv \, dx dy. \qquad (4.18)$$

From a computational point of view, the most useful weak form is the *Galerkin form*, derived from either (4.16) or (4.17) by k integrations by parts, i.e. Green's theorem. Possibly the simplest statement of Green's theorem in this context is that if $u, v \in \mathscr{H}^{(1)}(R)$ then

$$\int \cdots \int_R \left(u \frac{\partial v}{\partial x_i} + v \frac{\partial u}{\partial x_i} \right) dx = \int_{\partial R} uvn_i \, ds$$

where n_i is the component of the outward unit normal in the coordinate direction corresponding to the variable x_i. This Galerkin form has the minimum continuity requirements in the sense that neither Au nor A^*v need exist and u and v require the same degree of smoothness. It is written in the standard form as: find the function u from the set of admissible *trial functions* H_1 such that

$$a(u, v) = l(v) \quad \text{for all } v \in H_0 \qquad (4.19)$$

where H_0 is the space of admissible test functions, $a(,)$ is a *bilinear form*, and l is a *linear functional*.

For example, the Galerkin equation for a boundary value problem involving the differential operator

$$A = -\frac{\partial^2}{\partial x^2} - \frac{\partial^2}{\partial y^2} \qquad (4.20)$$

is based on the bilinear form

$$a_1(u, v) = \int\!\!\int_R \left[\left(\frac{\partial u}{\partial x} \right)\left(\frac{\partial v}{\partial x} \right) + \left(\frac{\partial u}{\partial y} \right)\left(\frac{\partial v}{\partial y} \right) \right] dx dy.$$

Using this notation, the equations (4.9) defining the Ritz approximation could be written as

$$\sum_j U_j a_1(\varphi_j, \varphi_i) = l(\varphi_i).$$

The linear functional l is based on the $\mathscr{L}_2$-inner product (f, v) of the inhomogeneous term and the test functions, but both the bilinear form and the linear functional can include boundary integrals. A *boundary value problem* based on

(4.20) is only equivalent to the weak form (4.19) if the trial space H_1 and the test space H_0 are chosen correctly. It can be shown that, if sufficient attention is given to the boundary conditions, a weak solution is unique and is also a classical solution. The existence and uniqueness of weak solutions is studied in Bers *et al.*, (1964), Nečas (1967) and Lions and Magenes (1972), but the only weak solution considered here is the Galerkin form. Following the definition of the Galerkin form (4.19), the Galerkin approximation U satisfies

$$a(U, V) = l(V) \qquad (4.21)$$

for all test functions $V \in K_N \subset \mathscr{H}_0$, where K_N is an N-dimensional subspace of the test space spanned by $\varphi_i (i = 1, \ldots, N)$, i.e. for any $V \in K_N$ there exist coefficients V_i such that

$$V = \sum_i V_i \varphi_i.$$

It then follows that an alternative formulation of (4.21) is

$$a(U, \varphi_i) = l(\varphi_i) \quad (i = 1, \ldots, N). \qquad (4.22)$$

Typically in the solution of boundary value problems the only difference between the test space and the trial space is the essential boundary conditions imposed on them. For example if

$$\partial R = \partial R_1 + \partial R_2$$

and the boundary condition on ∂R_2 is $u = h$, then

$$\mathscr{H}_1 = \{u : u \in \mathscr{H}^{(1)}(R) \quad \text{and } u = h \text{ on } \partial R_2\}$$

and

$$\mathscr{H}_0 = \{v : v \in \mathscr{H}^{(1)}(R) \quad \text{and } u = 0 \text{ on } \partial R_2\}.$$

Thus we can write

$$\mathscr{H}_1 = \mathscr{H}_0 \oplus \{w\}$$

where w is any function in $\mathscr{H}^{(1)}(R)$ such that $w = h$ on ∂R_2. The Galerkin equation can then be written as: find $u_0 \in \mathscr{H}_0$ such that

$$a(u_0, v) = l_0(v) \quad \text{for all } v \in \mathscr{H}_0 \qquad (4.23)$$

where

$$l_0(v) = l(v) - a(w, v). \qquad (4.24)$$

The problem is now written in terms of a bilinear form defined on $\mathscr{H}_0$ and it is possible to make use of results of Chapter 1 to prove existence and uniqueness of Galerkin solutions.

If $a(u, v)$ satisfies

$$a(u, u) \geqslant \gamma \|u\|^2 \quad \text{(for all } u \in \mathcal{H}_0) \tag{4.25}$$

then the Lax–Milgram lemma proves that a solution of (4.22) exists and is unique.

If the bilinear form satisfies the same condition for all $U \in K_N$ then the Galerkin approximation

$$U(\mathbf{x}) = \sum_i U_i \varphi_i(\mathbf{x}) \tag{4.26}$$

satisfies

$$GU = f$$

where

$$G = \{a(\varphi_j, \varphi_i)\}$$

is a *positive definite matrix*.

If the boundary conditions are inhomogeneous Dirichlet conditions, it is appropriate to use an approximation of the form

$$U(\mathbf{x}) = W(\mathbf{x}) + \sum_{i=1}^{N} U_i \varphi_i(\mathbf{x}) \tag{4.27}$$

where the function W satisfies the boundary conditions.

A formal proof of the equivalence of Galerkin solutions and classical solutions is not attempted here (see Showalter, 1977). It follows from Green's theorem that for a bilinear form $a(\ ,\ )$, there exists an operator A such that (Au, v) and $a(u, v)$ differ only by boundary terms. It is these boundary terms that determine the boundary value problem associated with a given weak form. The analysis is now illustrated by a simple example.

Define

$$a(u, v) = a_1(u, v) + a_2(u, v) \tag{4.28a}$$

where

$$a_1(u, v) = \iint_R \left\{ d_1 \left(\frac{\partial u}{\partial x} \right) \left(\frac{\partial v}{\partial x} \right) + d_2 \left(\frac{\partial u}{\partial y} \right) \left(\frac{\partial v}{\partial y} \right) \right\} dx$$

$d_1, d_2 \geqslant 0$ and

$$a_2(u, v) = \int_{\partial R_1} cuv \, ds$$

$c \geqslant 0$. Further assume that $u, v \in \mathcal{H}$ where

$$\mathcal{H} = \{w : w \in \mathcal{H}_2^{(1)}(R), w = 0 \text{ on } \partial R_2\}$$

with

$$\partial R_1 + \partial R_2 = \partial R.$$

Then define

$$l(v) = (f, v) + \langle g, v \rangle \tag{4.28b}$$

where

$$\langle g, v \rangle = \int_{\partial R_1} qv \, ds.$$

Let

$$A = -\frac{\partial}{\partial x} d_1 \frac{\partial}{\partial x} - \frac{\partial}{\partial y} d_2 \frac{\partial}{\partial y}$$

then from Green's theorem,

$$a(u, v) - (Au, v) = a_2(u, v) + a_3(u, v) \tag{4.29}$$

where in this example

$$a_3(u, v) = \int_{\partial R} \left(n_1 d_1 \frac{\partial u}{\partial x} + n_2 d_2 \frac{\partial u}{\partial y} \right) v \, ds$$

with $\mathbf{n} = (n_1, n_2)^{\mathrm{T}}$, the unit outward normal on ∂R.

The weak solution $u \in \mathcal{H}$ satisfies

$$a(u, v) = l(v) \quad \text{(for all } v \in \mathcal{H}) \tag{4.30}$$

where $a(\ ,\ )$ and l are defined by (4.28a) and (4.28b) respectively. Combining (4.29) and (4.30) leads to

$$(Au, v) - (f, v) = \langle g, v \rangle - a_2(u, v) - a_3(u, v) \quad \text{(for all } v \in \mathcal{H}). \tag{4.31}$$

It follows from (4.25) that the solution of (4.30) and hence of (4.31) is unique. The solution of the differential equation

$$Au - f = 0 \quad \mathbf{x} \in R \tag{4.32}$$

if it exists, would make the left-hand side of (4.31) zero. The right-hand side consists solely of boundary terms that can be written as

$$\langle g, v \rangle - \langle Lu, v \rangle \quad \text{for all } v \in \mathcal{H}$$

where in this example

$$\langle Lu, v \rangle = \int_{\partial R_1} \left(cu + n_1 d_1 \frac{\partial u}{\partial x} + n_2 d_2 \frac{\partial u}{\partial y} \right) v \, ds.$$

Note that $\int_{\partial R} = \int_{\partial R_1}$ since $v = 0$ on ∂R_2. Thus the left-hand side is zero if the boundary condition

$$Lu - g = 0 \quad \mathbf{x} \in \partial R_1 \tag{4.33a}$$

is satisfied in addition to the condition

$$u = 0 \quad \mathbf{x} \in \partial R_2 \tag{4.33b}$$

implicit in the definition of $\mathcal{H}$.

In this example the boundary condition (4.33a) becomes

$$cu + n_1 d_1 \frac{\partial u}{\partial x} + n_2 d_2 \frac{\partial u}{\partial y} - g = 0 \quad \mathbf{x} \in \partial R_1. \tag{4.34}$$

The boundary condition (4.33a) is the *natural boundary condition*, only satisfied by the solution of (4.30), whereas the *forced boundary condition* (4.33b) is imposed on all admissible functions $u \in \mathcal{H}$. The forced boundary conditions are invariably Dirichlet conditions whereas the natural boundary conditions contain derivatives of the same order as are found in the bilinear form.

To summarize: we have, from the Lax–Milgram lemma, that the solution to the weak form is unique and if a classical solution to the boundary value problem (4.32), (4.33a), and (4.33b) exists it is a solution of the weak form. Therefore, as we assume that a classical solution of the boundary value problem exists, the same function u is a solution of both problems.

If the Dirichlet conditions are inhomogeneous, say $u = h$ on ∂R_2, then an approximate solution of the form (4.27) is appropriate. The weak solution can be defined as before with

$$u = w + u_0$$

where $u_0 \in \mathcal{H}$ and w satisfies the inhomogeneous Dirichlet conditions. Then u_0 is the solution of

$$a(u_0, v) = l(v) \quad \text{(for all } v \in \mathcal{H}) \tag{4.35}$$

where now

$$l(v) = (f, v) - a(w, v) + \langle g, v \rangle. \tag{4.36}$$

Finite element approximations to the solution of such problems are frequently obtained by interpolating the boundary data. For example, assume that K_N denotes the space spanned by all φ_i regardless of boundary values and $\mathring{K}_N$ is the space spanned by those φ_i that are zero on ∂R_2. Then the approximate solution is

$$U = W + U_0$$

where

$$U_0 = \sum_{\varphi_i \in \mathring{K}_N} U_i \varphi_i$$

and

$$W = \sum_{\varphi_i \in K_N \setminus \mathring{K}_N} V_i \varphi_i$$

where clearly $K_N \backslash \overset{\circ}{K}_N$ contains only those φ_i capable of taking nonzero values on ∂R_2.

The parameters V_i are determined by the boundary data alone such that

$$U(\mathbf{x}_i) = V_i = u_\partial(\mathbf{x}_i)$$

for all nodes $\mathbf{x}_i \in \partial R_2$, where the boundary condition is $u = u_\partial$ on ∂R_2.

The parameters U_i are then determined from the Galerkin form (4.35) such that

$$a(U_0, \varphi_i) = (f, \varphi_i) - a(W, \varphi_i) + \langle g, \varphi_i \rangle \qquad (4.37)$$

for all $\varphi_i \in K_N$. In general this form of interpolation will not recover the boundary data exactly and so the method leads to an example of a *nonconforming* approximation in the sense that $U \in K_N \not\subset \mathscr{H}$.

There are a number of alternative methods of incorporating inhomogeneous boundary conditions. If it is considered necessary to recover the boundary conditions exactly, then for general nonconstant boundary data it is necessary to use *blending functions*. Then the approximation in elements bordering on the boundary is defined in terms of boundary data on the external side, together with nodal parameters on the internal sides. For example, in the configuration illustrated in Figure 4.1, the local approximation in element e can be written as

$$U = U_1 \frac{(x - (X + h))(y - (Y + h))}{h} \frac{}{h} - U_2 \frac{(x - X)(y - (Y + h))}{h} \frac{}{h} + u_\partial(x) \frac{(y - Y)}{h}$$

$$(4.38)$$

the two nodal parameters U_1 and U_2 to be determined from the Galerkin equations (4.35).

4.2.2 Numerical example 1

Consider the Galerkin approximation to the solution of

$$\frac{\partial^2 u}{\partial x^2} + \frac{\partial^2 u}{\partial y^2} - 2 = 0 \quad (-\tfrac{1}{2}\pi < x, y < \tfrac{1}{2}\pi) \qquad (4.39)$$

subject to

$$u(x, \pm\tfrac{1}{2}\pi) = 0 \quad (|x| \leqslant \tfrac{1}{2}\pi)$$

and   (4.39a)

$$u(\pm\tfrac{1}{2}\pi, y) = 0 \quad (|y| \leqslant \tfrac{1}{2}\pi).$$

A finite element approximation is required with piecewise bilinear functions. If the value of the solution is prescribed on the boundary (*Dirichlet boundary condition*) then this condition must be imposed on the trial space $\mathscr{H}$, whereas this is not necessary if the natural boundary conditions are given. It follows that in

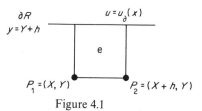

Figure 4.1

this problem it is necessary to restrict the trial space $\mathscr{H}$ to functions that satisfy the boundary condition $u = 0$ on the whole of the boundary.

The test functions also have to satisfy a homogeneous Dirichlet boundary condition on the whole boundary. Thus, in this example, the test space is the same as the trial space. The Galerkin solution therefore satisfies

$$a(U, \varphi) = (f, \varphi)$$

i.e.

$$\iint_R \left\{ \left(\frac{\partial U}{\partial x} \right) \left(\frac{\partial \varphi}{\partial x} \right) + \left(\frac{\partial U}{\partial y} \right) \left(\frac{\partial \varphi}{\partial y} \right) \right\} dxdy = \iint_R 2\varphi \, dxdy$$

where all the basis functions φ are zero on the whole of the boundary ∂R.

The approximating functions are defined on the region R, partitioned into $(T+1)^2$ square elements, by means of T equally spaced internal grid lines parallel to each axis (see Figure 1.8). Then the $N(=T^2)$ basis functions $\varphi_{ij}(x, y)$ $(i, j = 1, \ldots, T)$, for subspace K_N, are defined in Chapter 1 (Exercise 1.4) and clearly belong to $\mathscr{H}$ as they all vanish on the boundary.

The approximate solution is then of the form

$$U(x, y) = \sum_{i,j=1}^T U_{ij} \varphi_{ij}(x, y)$$

where U_{ij} is the approximate solution at the point (x_i, y_j) where $x_i = -\frac{1}{2}\pi + ih$ and $y_j = -\frac{1}{2}\pi + jh$ where $h = \pi/(T+1)$.

Thus the Galerkin equations become

$$\sum_{k,l=1}^T U_{kl} \iint_R \left\{ \left(\frac{\partial \varphi_{kl}}{\partial x} \right) \left(\frac{\partial \varphi_{ij}}{\partial x} \right) + \left(\frac{\partial \varphi_{kl}}{\partial y} \right) \left(\frac{\partial \varphi_{ij}}{\partial y} \right) \right\} dxdy$$

$$+ 2 \iint_R \varphi_{ij} dxdy = 0 \quad (i, j = 1, \ldots, T) \tag{4.40}$$

where $R = [-\frac{1}{2}\pi, \frac{1}{2}\pi] \times [-\frac{1}{2}\pi, \frac{1}{2}\pi]$.

As mentioned in Section 4.1, the equations (4.40) are assembled so that for each element e the element matrix G^e is constructed by transforming onto a reference element and integrating in the reference coordinates (p, q). Then the element matrix G^e is assembled into the global matrix A.

In this example, the transformation of an element

$$e = [X, X + h] \times [Y, Y + h] \quad (h = \pi/(T + 1))$$

onto $\varepsilon = [0, 1] \times [0, 1]$ is clearly

$$x = X + ph$$
$$y = Y + qh.$$

For this simple transformation, the Jacobian is $\partial(x, y)/\partial(p, q) = h^2$.

Following the notation of the previous chapter with a counter-clockwise local ordering of the nodes the element stiffness matrix is

$$G^e = \{a_{st}^e\} \quad 1 \leqslant s, t \leqslant n(= 4)$$

where e is an element with n nodes and where

$$a_{st}^e = \int_0^1 \int_0^1 \left\{ \left(\frac{\partial \varphi_s^e}{\partial p} \frac{\partial p}{\partial x} \right) \left(\frac{\partial \varphi_t^e}{\partial p} \frac{\partial p}{\partial x} \right) + \left(\frac{\partial \varphi_s^e}{\partial q} \frac{\partial q}{\partial y} \right) \left(\frac{\partial \varphi_t^e}{\partial q} \frac{\partial q}{\partial y} \right) \right\} \frac{\partial(x, y)}{\partial(p, q)} dp dq$$

since in this example

$$\frac{\partial p}{\partial y} = \frac{\partial q}{\partial x} = 0.$$

Thus

$$G^e = \frac{1}{6} \begin{bmatrix} 4 & -1 & -2 & -1 \\ -1 & 4 & -1 & -2 \\ -2 & -1 & 4 & -1 \\ -1 & -2 & -1 & 4 \end{bmatrix} \tag{4.41}$$

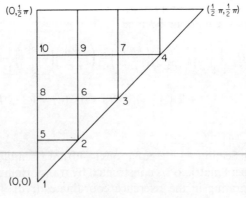

Figure 4.2

and the corresponding element vector to be added to the global right-hand side is

$$\mathbf{f}^e = -\frac{h^2}{2}(1, 1, 1, 1)^{\mathrm{T}}.$$

Each nodal equation thus has a contribution from four elements, the assembled equations being given in Exercise 4.2.

The approximate solution at the grid points shown in Figure 4.2 is given in Table 4.1; owing to symmetry only one-eighth of the region needs to be displayed. The theoretical solution is

$$u(x, y) = -(\tfrac{1}{2}\pi)^2 + x^2 + \frac{8}{\pi}\sum_{k=1}^{\infty}\frac{(-1)^{k+1}}{(2k-1)^3}\frac{\cosh(2k-1)y}{\cosh(2k-1)\tfrac{1}{2}\pi}\cos(2k-1)x.$$

If equation (4.39) is valid in the region $-\tfrac{1}{2}\pi < x, y < \tfrac{1}{2}\pi$ subject to the natural boundary conditions

$$\frac{\partial u(x, \pm\tfrac{1}{2}\pi)}{\partial y} = 0 \quad (|x| \leqslant \tfrac{1}{2}\pi)$$

and

$$\frac{\partial u(\pm\tfrac{1}{2}\pi, y)}{\partial x} = 0 \quad (|y| \leqslant \tfrac{1}{2}\pi)$$

the space of admissible functions, $\mathscr{H}$, also contains functions that take nonzero values on the boundary. The approximating subspace should also contain such functions, so we add basis functions $\varphi_{ij}(x, y)$ corresponding to the boundary points (x_i, y_j) where $x_i, y_j = \pm\tfrac{1}{2}\pi$. Such functions are nonzero in at most two elements rather than four elements. Note that as the mesh spacing is halved, the

Table 4.1

Point	Solution			
	$T = 3$	$T = 7$	$T = 15$	Exact
1	-1.534	-1.473	-1.459	-1.454
2		-1.321	-1.308	-1.304
3	-0.950	-0.907	-0.897	-0.894
4		-0.370	-0.362	-0.359
5		-1.394	-1.380	-1.376
6		-1.089	-1.078	-1.075
7		-0.566	-0.559	-0.556
8	-1.187	-1.146	-1.135	-1.132
9		-0.666	-0.660	-0.658
10		-0.698	-0.692	-0.690

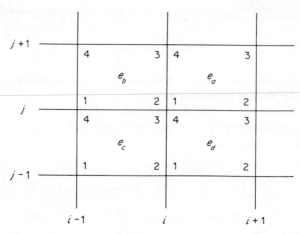

Figure 4.3 Rectangular grid with local numbering $(1, \ldots, 4)$ and global ordering (i, j) etc

nodal errors are divided by four, thus

$$|U_{ij} - u(x_i, y_j)| \approx O(h^2).$$

As each node is a vertex of four distinct elements, it follows that each nodal equation is assembled with contributions from four elements. Figure 4.3 shows that in general the node (i, j) is a vertex of four elements, labelled e_a, e_b, e_c and e_d, and is likely to have a different local number within each element. Thus the diagonal term in the global matrix, that is the coefficient of U_{ij} in the equation corresponding to φ_{ij}, can be written as

$$a_{11}^{e_a} + a_{22}^{e_b} + a_{33}^{e_c} + a_{44}^{e_d}$$

whereas the coefficient of $U_{i+1,j}$ in the same equation is

$$a_{12}^{e_a} + a_{43}^{e_d}$$

and the coefficient of $U_{i-1,j-1}$ is simply

$$a_{31}^{e_c}$$

and so on for the other six nodes with nonzero coefficients in the (i, j)-equation.
A straightforward combination of terms from (4.41) leads to

$$3U_{ij} - \frac{1}{3} \sum_{k=i-1}^{i+1} \sum_{l=j-1}^{j+1} U_{kl} + 2h^2 = 0 \tag{4.42a}$$

where $U_{kl} = 0$ if $k, l = 0, T+1$. These equations can be written as the $N \times N$ system

$$GU = f \tag{4.42b}$$

where

$$G = \begin{bmatrix} D & E & & & & \\ E & D & E & & & \\ & \cdot & \cdot & \cdot & & \\ & & \cdot & \cdot & \cdot & \\ & & & E & D & E \\ & & & & E & D \end{bmatrix}$$

with $T \times T$ blocks

$$D = \frac{1}{6}\begin{bmatrix} -8 & 1 & & & \\ 1 & -8 & 1 & & \\ & \cdot & \cdot & \cdot & \\ & & 1 & -8 & 1 \\ & & & 1 & -8 \end{bmatrix} \quad \text{and} \quad E = \frac{1}{6}\begin{bmatrix} 1 & 1 & & & \\ 1 & 1 & 1 & & \\ & \cdot & \cdot & \cdot & \\ & & 1 & 1 & 1 \\ & & & 1 & 1 \end{bmatrix}$$

and $\mathbf{f} = 2h^2(1, 1, \ldots, 1)^{\mathrm{T}}$.

As the basis functions φ_{ij} are tensor products of one-dimensional functions, it is possible to write the Galerkin equations as the product of one-dimensional forms, i.e.

$$\{\delta_x^2 I_y + \delta_y^2 I_x\} U_{ij} - 2h^2 = 0 \quad (i, j = 1, \ldots, T) \tag{4.42c}$$

where δ_x^2, δ_y^2 are second-central-difference operators and I_x, I_y are 'Simpson's Rule' operators defined by

$$I_x U_{ij} = \tfrac{1}{6}[U_{i-1\,j} + 4U_{ij} + U_{i+1\,j}]$$

and similarly for I_y.

The form (4.42c) leads directly to the representation of the stiffness matrix as the tensor product

$$G = G_x \otimes B_y + B_x \otimes G_y$$

where

$$B_x = B_y = \frac{h}{6}\begin{bmatrix} 4 & 1 & & & & \\ 1 & 4 & 1 & & & \\ & \cdot & \cdot & \cdot & & \\ & & \cdot & \cdot & \cdot & \\ & & & 1 & 4 & 1 \\ & & & & 1 & 4 \end{bmatrix}$$

$$G_x = G_y = \frac{1}{h}\begin{bmatrix} 2 & -1 & & & & \\ -1 & 2 & -1 & & & \\ & \cdot & \cdot & \cdot & & \\ & & \cdot & \cdot & \cdot & \\ & & & -1 & 2 & -1 \\ & & & & -1 & 2 \end{bmatrix}.$$

A definition of matrix tensor products, and the *alternating-direction-Galerkin* methods that make use of them to speed the solution, can be found in Section 4.5.3.

Exercise 4.2 (a) Verify the formulae (4.42a, b, c).
(b) Using the Galerkin method with piecewise bilinear basis functions, calculate the coefficients of the system (4.40) corresponding to the basis function φ_{ij} when x_i, y_j is (i) a corner point and (ii) a side point, assuming in both cases that natural boundary conditions are specified.

Exercise 4.3 Compute the element stiffness matrix G^e for a bilinear element $[X, X + h] \times [Y, Y + k]$ $(k \neq h)$ corresponding to the Galerkin equation (4.40).

4.2.3 Numerical example 2

Consider a Galerkin approximation to the solution of

$$\frac{\partial^2 u}{\partial x^2} + \frac{\partial^2 u}{\partial y^2} + 2\alpha \frac{\partial u}{\partial y} = 0 \tag{4.43}$$

where $-\frac{1}{2}\pi \leqslant x, y \leqslant \frac{1}{2}\pi$ and α is a parameter.
   Assume that the boundary conditions are

$$u(\pm \tfrac{1}{2}\pi, y) = 0 \quad (|y| \leqslant \tfrac{1}{2}\pi)$$
$$u(x, -\tfrac{1}{2}\pi) = 0 \quad (|x| \leqslant \tfrac{1}{2}\pi) \tag{4.44}$$

and

$$u(x, \tfrac{1}{2}\pi) = (\tfrac{1}{2}\pi)^2 - x^2 \quad (|x| \leqslant \tfrac{1}{2}\pi)$$

Then the Galerkin finite element method can be applied to this problem to derive an approximate solution in terms of the piecewise bilinear functions employed earlier. We introduce the function

$$w(x, y) = [(\tfrac{1}{2}\pi)^2 - x^2](\tfrac{1}{2}\pi + y)\frac{1}{\pi} \tag{4.45}$$

which satisfies the boundary conditions, and then we define a Galerkin approximation of the form

$$U(x, y) = \sum_{i,j=1}^{T} U_{ij}\varphi_{ij}(x, y) + W(x, y)$$

by means of (4.37).
   The boundary data matching can be achieved using several alternatives; (4.45) is just one such possibility. The function $W(x, y) = w(x, y)$ defined by (4.45) can be thought of as a function blended over the whole region as it only vanishes on $y = -\frac{1}{2}\pi$.

Following (4.38) it would be possible to blend over one column of elements only, i.e. $y\in[\frac{1}{2}\pi - h, \frac{1}{2}\pi]$, in which case

$$W(x,y) = \begin{cases} [(\tfrac{1}{2}\pi)^2 - x^2](h - \tfrac{1}{2}\pi + y)\dfrac{1}{h} & y\in[\tfrac{1}{2}\pi - h, \tfrac{1}{2}\pi] \\ 0 & y < \tfrac{1}{2}\pi - h. \end{cases} \tag{4.46}$$

In this case only the matrix components corresponding to nodes in elements bordering the boundary $y = \frac{1}{2}\pi$ are affected by the term $a(W, \varphi_{ij})$ from (4.37), i.e. nodes $(x_i, y_j), y_j = \frac{1}{2}\pi - h$.

The third alternative is to interpolate the boundary conditions; in this case the approximate solution is

$$U(x,y) = \sum_{i,j=1}^{T} U_{ij}\varphi_{ij}(x,y) + \sum_{i=1}^{T} U_{i,T+1}\varphi_{i,T+1}(x,y) \tag{4.47}$$

the functions $\varphi_{i,T+1}(x,y)$ are nonzero on the boundary $y = \frac{1}{2}\pi$ and

$$U_{i,T+1} = (\tfrac{1}{2}\pi)^2 - x_i^2. \tag{4.47a}$$

Thus

$$W(x,y) = \sum_{i=1}^{T} U_{i,T+1}\varphi_{i,T+1}(x,y) \tag{4.48}$$

and the remaining coefficients are determined by the Galerkin equations (4.37).

For the differential equation (4.43) the Galerkin weak form is

$$a(u,v) = \iint_R \left\{ \left(\frac{\partial u}{\partial x}\right)\left(\frac{\partial v}{\partial x}\right) + \left(\frac{\partial u}{\partial y}\right)\left(\frac{\partial v}{\partial y}\right) - v\frac{\partial u}{\partial y} + u\frac{\partial v}{\partial y} \right\} dxdy = 0 \tag{4.49}$$

or equivalently

$$a(u_0, v) = -a(w, v).$$

As all the boundary conditions are Dirichlet, the space of admissible functions $\mathscr{H}$ includes only u_0 and v that vanish on ∂R the boundary of

$$R = [-\tfrac{1}{2}\pi, \tfrac{1}{2}\pi] \times [-\tfrac{1}{2}\pi, \tfrac{1}{2}\pi].$$

The results of the computation, using (4.45) and the various grids described earlier, are given in Table 4.2. Figure 4.4 illustrates the position of the points referred to in the table. Although there is not the same degree of symmetry as before, it is still only necessary to consider half of the region. The theoretical solution of equation (4.43) subject to the boundary conditions is

$$u(x,y) = \frac{4}{\pi}e^{\alpha(\pi/2 - y)} \sum_{k=1}^{\infty} \frac{\cos(2k-1)x}{(2k-1)^3}\{A\sinh(\theta_k y) + B\cosh(\theta_k y)\} \tag{4.50}$$

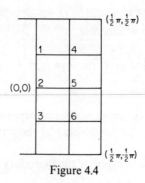

Figure 4.4

where

$$A = (\sinh(\theta_k \tfrac{1}{2}\pi))^{-1}$$
$$B = (\cosh(\theta_k \tfrac{1}{2}\pi))^{-1}$$

and

$$\theta_k = ((2k - 1)^2 + \alpha^2)^{1/2}$$

with $\alpha = 1$ in this example.

Table 4.2 Boundary data blended across the whole region

Point	Solution			
	$T = 3$	$T = 7$	$T = 15$	Exact
1	1.826	1.822	1.821	1.821
2	1.324	1.314	1.311	1.310
3	0.934	0.870	0.859	0.855
4	1.301	1.309	1.310	1.311
5	0.940	0.933	0.931	0.931
6	0.665	0.617	0.608	0.605

Table 4.3 Boundary data blended across only elements adjacent to boundary

Point	Solution			
	$T = 3$	$T = 7$	$T = 15$	Exact
1	1.897	1.839	1.826	1.821
2	1.359	1.322	1.313	1.310
3	0.937	0.871	0.859	0.855
4	1.364	1.323	1.314	1.311
5	0.962	0.938	0.933	0.931
6	0.663	0.617	0.608	0.605

Table 4.4 Boundary data interpolated at boundary nodes

Point	Solution			
	$T = 3$	$T = 7$	$T = 15$	Exact
1	1.804	1.817	1.820	1.821
2	1.286	1.305	1.309	1.310
3	0.886	0.860	0.856	0.855
4	1.282	1.305	1.309	1.311
5	0.909	0.926	0.930	0.931
6	0.627	0.608	0.606	0.605

The results using (4.46) are given in Table 4.3 and those for (4.47) in Table 4.4. Tables 4.2, 4.3, and 4.4 all indicate nodal errors decreasing as $O(h^2)$.

Leaving the position of the nodes unaltered, it is possible to replace each bilinear square element by two linear triangular elements. Following an analysis analogous to that outlined in numerical example 1, each triangle can be mapped onto a reference element which in this case would be half a unit square. The same linear transformation can be used in this case and the element stiffness matrix corresponding to the

$$\frac{\partial^2 u}{\partial x^2} + \frac{\partial^2 u}{\partial y^2}$$

term can be written as

$$A^e = \frac{1}{2}\begin{bmatrix} 1 & \cdot & -1 \\ \cdot & 1 & -1 \\ -1 & -1 & 2 \end{bmatrix} \tag{4.51}$$

for the element shown in Figure 4.6(a). If the two elements shown in Figure 4.6(b) are taken together the 'element' matrix corresponding to this two-element patch is

$$A^e = \frac{1}{2}\begin{bmatrix} 2 & -1 & \cdot & -1 \\ -1 & 2 & -1 & \cdot \\ \cdot & -1 & 2 & -1 \\ -1 & \cdot & -1 & 2 \end{bmatrix}$$

and this can be compared directly with the bilinear element matrix given in numerical example 1.

If symmetry is to be employed to reduce the computation, it is necessary to use a grid such as is shown in Figure 4.5. The results of the numerical solution of (4.43) subject to (4.44) using linear elements with the same nodal positions as in the bilinear example are given in Table 4.5. The boundary condition on $y = \frac{1}{2}\pi$ is interpolated in terms of the basis functions. These results should be compared

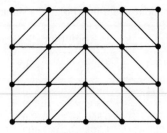

Figure 4.5

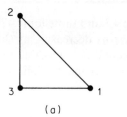 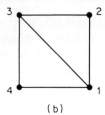

Figure 4.6

Table 4.5 Linear triangular elements

Point	Solution			
	$T = 3$	$T = 7$	$T = 15$	Exact
1	1.798	1.814	1.819	1.821
2	1.295	1.305	1.309	1.310
3	0.871	0.859	0.856	0.855
4	1.309	1.310	1.310	1.311
5	0.926	0.929	0.930	0.931
6	0.598	0.604	0.605	0.605

with Table 4.4: the differences are not significant *in this example*, but this is not always true. The convergence of the nodal values again appears to be $O(h^2)$.

4.2.4 Nonconforming methods (see also Section 6.4.5)

In the Galerkin method, a finite element implementation assumes that the integral over the entire region R can be split into the sum of integrals over individual elements. This is how the global matrix is assembled, by adding together the separate element matrices.

For the equality

$$\iint_R f(\mathbf{x})\,d\mathbf{x} = \sum_e \iint_e f(\mathbf{x})\,d\mathbf{x}$$

to hold, it is necessary for f to have at most finite jumps across the element boundaries. Thus, if $u(\mathbf{x})$ is continuous but has a discontinuous normal derivative then an integrand such as

$$u_x^2 + u_y^2$$

will have a finite jump across the boundary between two elements. On the other hand if $u(\mathbf{x})$ itself has a discontinuity, then the derivative has to be defined in terms of *Dirac delta functions*. In one dimension, if u is the *Heaviside function* $H(x - x_0)$ then

$$u(x) = \begin{cases} 0 & x < x_0 \\ 1 & x \geqslant x_0 \end{cases}$$

then the derivative can be defined as the delta function, i.e.

$$\frac{du}{dx} = \delta(x - x_0)$$

where

$$\delta(t) = 0 \quad t \neq 0$$

and

$$\int_{-\infty}^{\infty} \delta(t)\,dt = 1.$$

Clearly if $\iint_R$ is replaced by $\sum_e \iint_e$ then such boundary delta functions are lost. Nonconforming elements lack the continuity necessary to be admissible functions in the classical Ritz or Galerkin formulations. Because of discontinuities along the element boundaries they lead to integrands $f(\mathbf{x})$ for which

$$\iint_R f(\mathbf{x})\,d\mathbf{x} \neq \sum_e \iint_e f(\mathbf{x})\,d\mathbf{x}.$$

It is $\iint_R$ that contains the delta functions and $\sum_e \iint_e$ that is used to construct the global matrix.

One example of a nonconforming approximation that *sometimes* does give acceptable results is the linear triangle with nodes only at the midsides (see Figure 4.7).

Consider the solution of the numerical example 1 given earlier in Section 4.2.2. On a uniform grid of right-angle triangles, a mesh of nonconforming (de

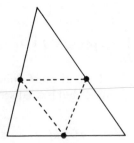

Figure 4.7

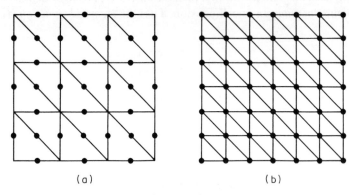

(a)   (b)

Figure 4.8

Veubeke) elements has almost as many degrees of freedom (i.e. nodes) as a finer mesh of conforming triangular or rectangular elements (see Figure 4.8).

With $T = 15\,(h = \pi/32)$, the conforming approximation on $[0, \tfrac{1}{2}\pi] \times [0, \tfrac{1}{2}\pi]$ has $(\tfrac{1}{2}(T + 1))^2 = 64$ nodes, whereas the nonconforming approximation on the same tessellation has 192 nodes. With this mesh, the values obtained at nodes numbered $1, \ldots, 14$ in Figure 4.9 are given in Table 4.6. The node numbers $1, 4, 5, 8, 9, 12,$ and 13 are duplicated because of the symmetry in both the numerical and analytical solution.

The values of the vertices I, II, III, and IV in Figure 4.9 are found from the nodal values using linear interpolation as the elements are linear. The accuracy obtained in this example does not appear to differ significantly from that expected with a comparable conforming approximation (see Table 4.1), but such comparability is not universal.

An alternative comparison of conforming and nonconforming triangular elements is illustrated by the numerical solution of (4.43) subject to (4.44), i.e. numerical example 2. Figure 4.10 and 4.11 are respectively the conforming and nonconforming approximations with $T = 7$. The conforming grid has 28 nodes

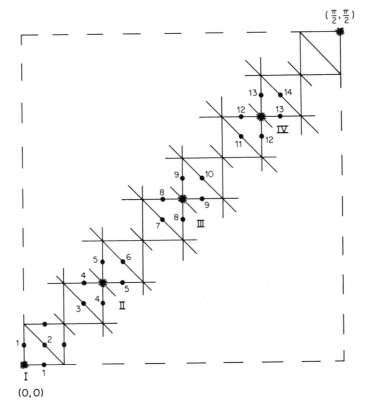

Figure 4.9

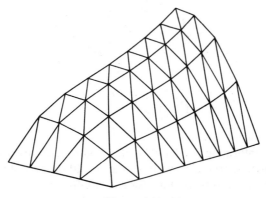

Figure 4.10

Table 4.6 Linear nonconforming elements

Nodes	Nodal Value	Vertex	Vertex value Approx.	Vertex value Exact
1	− 1.450 ⎫	I	− 1.457	− 1.454
2	− 1.443 ⎭			
3	− 1.368 ⎫	II	− 1.304 ⎫	
4	− 1.336 ⎭		⎬	− 1.304
5	− 1.263 ⎫	II	− 1.305 ⎭	
6	− 1.221 ⎭			
7	− 1.015 ⎫	III	− 0.891 ⎫	
8	− 0.953 ⎭		⎬	− 0.894
9	− 0.828 ⎫	III	− 0.891 ⎭	
10	− 0.765 ⎭			
11	− 0.494 ⎫	IV	− 0.350 ⎫	
12	− 0.422 ⎭		⎬	− 0.359
13	− 0.292 ⎫	IV	− 0.350 ⎭	
14	− 0.235 ⎭			

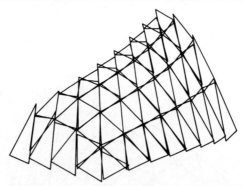

Figure 4.11

whereas the nonconforming grid has 92 nodes. Figure 4.12 illustrates the solution
with 6-node quadratic triangles, with a total of 120 degrees of freedom. These
figures should be viewed as three-dimensional wire-frame surfaces with the height
of the wire frame corresponding to the value of the solution along the element

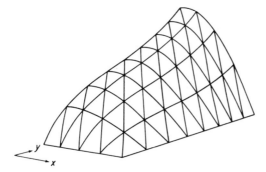

Figure 4.12

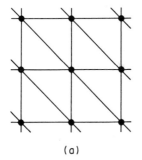

(a)

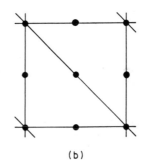

(b)

Figure 4.13

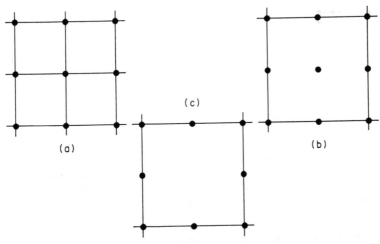

(a)

(b)

(c)

Figure 4.14

boundaries. The surface is then viewed from a point with (x, y) coordinates approximately $(\frac{3}{4}\pi, -\frac{3}{4}\pi)$.

4.2.5 Higher order elements

It is possible to solve numerical example 2 with the same number of nodes but fewer elements if higher order shape functions are used. Thus, for example, we can use the formulation of Figure 4.13(b) rather than Figure 4.13(a) or Figure 4.14(b) or (c) in place of Figure 4.14(a). In fact the 8-node element in Figure 4.14(c) involves fewer nodes than Figure 4.14(a). Thus the nodal values given in Table 4.7 correspond to a smaller algebraic system than the alternative 9-node or 4-node elements. The nodal numbers in Table 4.7 relate to Figure 4.4 as with the earlier solution. For each value of T, the number of degrees of freedom for 6-node triangles and 9-node $(2h \times 2h)$ squares are identical to that for the 3- and 4-node grids illustrated in Figure 4.5 and Figure 4.4 respectively.

The apparent rate of convergence of the largest errors is (very approximately) the $O(h^3)$ expected from the theory (see Section 6.3.3). Figure 4.12 illustrates the solution as a 3-dimensional perspective plot of the $T = 15$ grid of 6-node quadratic triangles.

Table 4.7 Higher order elements

Points	9-node biquadratic		8-node biquadratic		6-node quadratic		Exact
	$T = 7$	$T = 15$	$T = 7$	$T = 15$	$T = 7$	$T = 15$	
1	1.82111	1.82074	1.81840	1.82071	1.82226	1.82089	1.82071
2	1.30924	1.30992	1.30989	1.30992	1.30986	1.30998	1.30995
3	0.85155	0.85488	0.84810	0.85483	0.84954	0.85455	0.85506
4	1.31064	1.31072	1.31591	1.31081	1.30967	1.31063	1.31070
5	0.93003	0.93066	0.93063	0.93070	0.93025	0.93064	0.93070
6	0.60289	0.60529	0.60051	0.60527	0.60795	0.60553	0.60543

Exercise 4.4 Verify that if (p_i, q_i), $i = 1, \ldots, k$, are the nodes on the reference element and if $p^{s_j}q^{t_j}$, $j = 1, \ldots, k$, are the terms in the basis functions, then the coefficient of $p^{s_i}q^{t_i}$ in the basis function φ_j is $(D^{-1})_{ij}$ where the matrix D is defined as

$$D_{ij} = p_i^{s_j}q_i^{t_j} \quad i, j = 1, \ldots, k.$$

Thus the columns of D^{-1} give the basis functions on the reference element. This is how the basis function could be computed in a practical program, rather than coding in all the separate coefficients given in Chapter 3. Given the matrix D^{-1} and (x_i, y_i), $i = 1, \ldots, k$, the positions of the nodes in the physical element, derive an expression for $\partial x/\partial p$, $\partial x/\partial q$, etc. and hence for $\partial(x, y)/\partial(p, q)$.

Exercise 4.5 Calculate a typical nonzero component $a(W, \varphi_{i, T})$ for the boundary function $W(x, y)$ given by (a) (4.46) and (b) (4.48).

Exercise 4.6 Given the weak form (4.49), verify that the natural boundary conditions are

$$\frac{\partial u}{\partial x} = 0 \quad |x| = \tfrac{1}{2}\pi$$

and

$$\frac{\partial u}{\partial y} + u = 0 \quad |y| = \tfrac{1}{2}\pi.$$

Verify that a conforming approximation based on the alternative weak form

$$\iint \left\{ \left(\frac{\partial u}{\partial x}\right)\left(\frac{\partial v}{\partial x}\right) + \left(\frac{\partial u}{\partial y}\right)\left(\frac{\partial v}{\partial y}\right) - 2v\frac{\partial u}{\partial y} \right\} dxdy = 0$$

leads to *identical* Galerkin equations for the *Dirichlet* problem.

4.2.6 Penalty methods: numerical example 3

One further method of dealing with inhomogeneous boundary data is to modify the functional (Bramble and Schatz, 1970; Babuška, 1973; Nitsche, 1971). The definition of a weak solution given by (4.30) involves boundary integrals $a_2(u, v)$ and $\langle g, v \rangle$. These additional terms can be followed through to the modified boundary condition (4.34).

If for example

$$a_2(u, v) = \frac{c}{h} \int_{\partial R} uv \, ds$$

and we replace $\langle g, v \rangle$ by $c\langle g, v \rangle/h$ it follows that the natural boundary condition is modified by the term

$$\frac{c}{h}(u - g).$$

Thus as h tends to zero, this term tends to dominate and the natural boundary condition tends to

$$u - g = 0.$$

Numerical example 3

Consider again the numerical solution of (4.43) subject to (4.44). This time a Galerkin solution is required using the penalty approach outlined above, using the same grid and the same bilinear basis functions as before. This method replaces the inhomogeneous Dirichlet condition by a natural boundary condition, so it is necessary to incorporate basis functions that can take nonzero values on the boundary $y = \frac{1}{2}\pi$, i.e.

$$U = \sum_{i=1}^{T} \sum_{j=1}^{T+1} U_{ij}\varphi_{ij}(x, y). \tag{4.52}$$

The weak form to be used is then

$$\iint_R \left\{ \left(\frac{\partial u}{\partial x}\right)\left(\frac{\partial v}{\partial x}\right) + \left(\frac{\partial u}{\partial y}\right)\left(\frac{\partial v}{\partial y}\right) + u\frac{\partial v}{\partial y} - v\frac{\partial u}{\partial y} \right\} dxdy + \frac{c}{h}\int_{\partial R} ux\,dx$$

$$= \frac{c}{h}\int_{\partial R_1} gv\,dx \quad \text{for all } v \in \mathcal{H} \tag{4.53}$$

i.e. $\qquad\qquad\qquad a(u, v) = \langle g, v \rangle$

where ∂R_1 is the boundary segment $y = \frac{1}{2}\pi$, $|x| \leqslant \frac{1}{2}\pi$. The space $\mathcal{H}$ in this case consists of functions that are zero on the other three sides. Following (4.37), the nodal parameters in the Galerkin approximation (4.52) are defined by the equations

$$a(u, \varphi_{ij}) = \langle g, \varphi_{ij} \rangle \quad \begin{cases} i = 1, \ldots, T \\ j = 1, \ldots, T+1. \end{cases}$$

Thus the Galerkin equations are

$$\sum_{k=1}^{T} \sum_{l=1}^{T+1} U_{kl} \left\{ \int_{-\pi/2}^{\pi/2} \int_{-\pi/2}^{\pi/2} \left[\left(\frac{\partial \varphi_{ij}}{\partial x}\right)\left(\frac{\partial \varphi_{kl}}{\partial x}\right) + \left(\frac{\partial \varphi_{ij}}{\partial y}\right)\left(\frac{\partial \varphi_{kl}}{\partial y}\right) \right. \right.$$

$$\left. \left. + \varphi_{kl}\left(\frac{\partial \varphi_{ij}}{\partial y}\right) - \varphi_{ij}\left(\frac{\partial \varphi_{kl}}{\partial y}\right) \right] dxdy + \frac{c}{h}\int_{-\pi/2}^{\pi/2} \varphi_{ij}(x, \tfrac{1}{2}\pi)\varphi_{kl}(x, \tfrac{1}{2}\pi)dx \right\}.$$

$$= \frac{c}{h}\int_{-\pi/2}^{\pi/2} ((\tfrac{1}{2}\pi)^2 - x^2)\varphi_{ij}(x, \tfrac{1}{2}\pi)\,dx \quad \begin{cases} i = 1, \ldots, T \\ j = 1, \ldots, T+1. \end{cases}$$

Clearly the two boundary integral terms are only nonzero for $j = T+1$. The results for different values of c, using the same grids as in the previous examples, are given in Tables 4.8 and 4.9.

Points 7 and 8 are the boundary grid points in Figure 4.4 above internal nodes 1 and 4 respectively. A convergence rate of $O(h^2)$ can be identified at the internal nodes of Tables 4.8 and 4.9. An alternative weak form for this problem is

$$\iint_R \left\{ \left(\frac{\partial u}{\partial x}\right)\left(\frac{\partial v}{\partial x}\right) + \left(\frac{\partial u}{\partial y}\right)\left(\frac{\partial v}{\partial y}\right) - 2\frac{\partial u}{\partial y}v \right\} dxdy + \frac{c}{h}\int_{\partial R_1} (u - g)v\,dx = 0.$$

$$\tag{4.54}$$

It is easily verified (Exercise 4.7) that the natural boundary conditions of (4.53) and (4.54) are different; hence, while both methods converge to the analytic

Table 4.8 Boundary data incorporated using a penalty functional ($c = 10$)

Point	Solution			
	$T = 3$	$T = 7$	$T = 15$	Exact
1	1.706	1.743	1.776	1.821
2	1.216	1.252	1.277	1.310
3	0.839	0.824	0.836	0.855
4	1.214	1.252	1.278	1.311
5	0.861	0.888	0.907	0.931
6	0.593	0.584	0.592	0.605
7	2.315	2.366	2.409	2.467
8	1.765	1.774	1.806	1.851

Table 4.9 Boundary data incorporated using a penalty functional ($c = 100$)

Point	Solution			
	$T = 3$	$T = 7$	$T = 15$	Exact
1	1.878	1.829	1.820	1.821
2	1.340	1.314	1.309	1.310
3	0.924	0.866	0.856	0.855
4	1.338	1.315	1.310	1.311
5	0.948	0.933	0.930	0.931
6	0.653	0.613	0.606	0.605
7	2.530	2.479	2.467	2.467
8	1.958	1.864	1.852	1.851

solution as h tends to zero, for a fixed value of c/h they give slightly different numerical values.

Exercise 4.7 Verify that the natural boundary condition on ∂R_1 corresponding to (4.53) is

$$\frac{\partial u}{\partial y} + u + \frac{c}{h}(u - g) = 0$$

and corresponding to (4.54) is

$$\frac{\partial u}{\partial y} + \frac{c}{h}(u - g) = 0.$$

4.3 Petrov–Galerkin Methods

In the Galerkin equations, it is natural to assume that the approximation U and the test functions V are defined in terms of the same set of basis functions

$\varphi_i(i = 1, \ldots, N)$. By so doing, in problems for which both Ritz and Galerkin approximations can be defined the two methods lead to alternative formulations of the same solution. A different approximation, in terms of the same basis functions $\varphi_i \in K_N$, is obtained if the test functions are defined in terms of some $\psi_i \in L_N (i = 1, \ldots, N)$, where K_N and L_N are different subspaces of $\mathcal{H}$.

The use of two sets, $\{\varphi_i\}$ and $\{\psi_i\}$, of dissimilar basis functions to define an approximation of the form

$$a(U, \psi_i) = (f, \psi_i) \quad (i = 1, \ldots, N) \tag{4.55}$$

has led various authors to develop so-called *conjugate approximations* (Oden, 1972, and references therein) and the so-called *method of weighted residuals* (Finlayson and Scriven, 1967, and references therein).

The basic extension of the Galerkin method is due to G. I. Petrov (in 1946) and the name *Petrov–Galerkin* is frequently applied to such methods when used in a finite element context (Anderssen and Mitchell, 1979).

In this section, we consider a number of different implementations; for example, *upwinding* is the generic name given to methods that lead to nonsymmetric or noncentred Galerkin equations. Collocation can be viewed as a Petrov–Galerkin method in which the test functions are Dirac delta functions, while collocation–Galerkin takes different forms of test functions corresponding to different nodes. In the H^{-1} and H^1 Galerkin methods and in the method of least squares, the form of the inner product is also changed.

Other examples are given in Chapter 7, where it is found convenient to have test and trial functions of different degrees.

4.3.1 Upwinding

As a motivation for studying such methods, consider the differential equation

$$\frac{\partial^2 u}{\partial x^2} + \frac{\partial^2 u}{\partial y^2} + 2\alpha \frac{\partial u}{\partial y} = 0. \tag{4.56}$$

That is, an equation similar to that studied in the preceding numerical examples but with $\alpha \gg 1$, i.e. a first derivative term that may be arbitrarily large. The boundary conditions are as before and the same grids and bilinear basis functions are used. The results are probably best illustrated by considering a graph of the solution along the line $x = 0$. In the previous examples ($\alpha = 1$) this would have shown the rapid and smooth convergence of the numerical solutions to the exact solution as the grid spacing, h, tends to zero. However, with the larger coefficient ($\alpha \gg 1$) of the first derivative term the picture is altogether different.

Figure 4.15 illustrates the behaviour of the solution along $x = 0$ for different values of α. The analytic solution is given by (4.50) and the values of T refer to the meshes as used in the earlier examples. As the value of x increases so the solution of the differential equation (4.56) exhibits a characteristic form. Over most of the

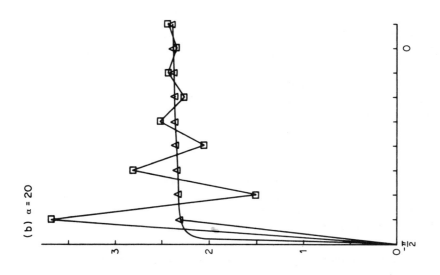

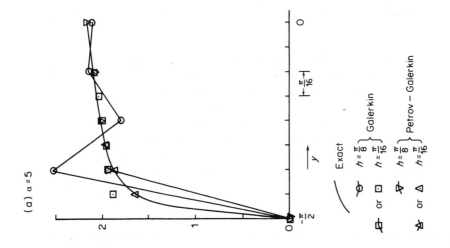

Figure 4.15

region, i.e. excluding a small strip $y \in [-\frac{1}{2}\pi, -\frac{1}{2}\pi + \varepsilon]$, where ε tends to zero as α increases, the solution is very close to the boundary value at $y = \frac{1}{2}\pi$, thus

$$u(x, y) \approx u(x, \tfrac{1}{2}\pi) \quad y > -\tfrac{1}{2}\pi + \varepsilon.$$

All the variation from the boundary value at $y = \frac{1}{2}\pi$ occurs in the thin strip of width ε. This is an example of a *boundary layer* effect. Equation (4.56) can be interpreted as a *diffusion* term, $\partial^2 u/\partial x^2 + \partial^2 u/\partial y^2$, together with *convection* by a flow with velocity -2α parallel to the y-axis. Given this interpretation of (4.56) as an example of a *convection–diffusion equation*, the flow is dominated by the convection and there is a boundary layer at the outflow boundary.

The numerical solution illustrated in Figure 4.15 exhibits two standard characteristics of such problems:

(a) If the product of the grid size and the flow speed (i.e. $h\alpha$) is too large then there are oscillations in the solution. These oscillations are greatest at the boundary with the boundary layer.
(b) For a given flow speed α, the oscillations will disappear if the grid size h is reduced to a level at which the approximation can resolve the solution accurately within the boundary layer (see Gresho *et al.*, 1978).

As problems in which $\alpha > 100$ are not unrealistic, it is clear from Figure 4.15 that a sufficiently fine *uniform* mesh could prove prohibitively expensive in terms of computer time and storage. Thus there are at least two possible policies to adopt:

(a) Construct a method that provides acceptable accuracy on a coarse grid.
(b) Refine the grid only in the region of the boundary layer and thus retain the same overall number of nodes in the approximation.

A combination of both may be necessary *in extremis*. In either case the idea is to construct accurate solutions that are free from the spurious oscillations. It is important to remember that they are *spurious* and that care should be exercised when using a 'wiggle suppressant' so that physically meaningful oscillations are not damped unnaturally.

It is this property of damping spurious wiggles that has led to the adoption of Petrov–Galerkin methods with different test and trial functions. As these methods often lead to the introduction of numerical dissipation of energy, it is important that their use should be carefully controlled. Another viewpoint is that Galerkin methods naturally lead to central difference formulae in terms of the nodal parameters. Petrov–Galerkin methods provide an opportunity to include backward or forward differences: hence the name *upwinding* (or *downwinding*) of the Galerkin equations for flow equations by biasing upstream (or downstream) relative to the flow direction.

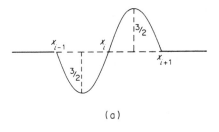

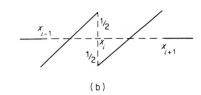

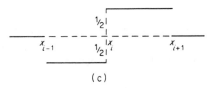

Figure 4.16 Perturbations for piecewise linear basis functions: (*a*) piecewise quadratic; (*b*) piecewise linear; (*c*) piecewise constant

Analysis in one dimension

It is possible in one dimension to use as trial functions φ_i, and as test functions

$$\psi_i(x) = \varphi_i(x) + \gamma_i \sigma_i(x) \qquad (4.57)$$

where σ_i can be any one of the three functions illustrated in Figure 4.16. Note that both the piecewise linear and piecewise constant forms are discontinuous. The coefficients in the weak form are well defined since it is understood that

$$a(U, \psi_i) = (f, \psi_i) \qquad (4.58)$$

is interpreted as

$$\sum_e a^e(U^e, \psi_i^e) = (f, \psi_i^e)$$

where superscript *e* denotes the restriction to an individual element as in Section 4.1. Thus it is not necessary to integrate across element boundaries and so the integrals remain finite. Later in this chapter approximations using discontinuous trial functions are considered; such approximations are known as nonconforming. Note that all the test functions chosen are such that

$$\sum_e \int \psi_i^e = \sum_e \int \varphi_i^e$$

so that the relative scaling of the equations is unaltered.

An analysis of the one-dimensional problem

$$\frac{d^2 u}{dx^2} + k\frac{du}{dx} = 0 \tag{4.59}$$

shows that for the piecewise quadratic or piecewise constant forms,

$$\gamma_i = \coth\left(\frac{kh}{2}\right) - \frac{2}{kh} \tag{4.60}$$

leads to exact recovery of the analytic solution

$$u(x) = e^{-kh} \tag{4.61}$$

at the nodes. This is an example of nodal *superconvergence* when the numerical solution at the nodes is much more accurate than at intermediate points. If the piecewise linear form is used, the parameters

$$\gamma_i = \frac{kh}{2}\coth\left(\frac{kh}{2}\right) - 1 \tag{4.62}$$

lead to the same superconvergence.

The parameter $P_e = kh/2$ is known as the *cell Peclet* number. Details of higher order approximations to the model problem (4.59) can be found in Mitchell and Griffiths (1980).

Two-dimensional problems

In two dimensions, the connection between the coefficient of the first derivative term and the exponent in the analytic solution is not as straightforward as that indicated by (4.59) and (4.61). Consider, for example, the solution of (4.56) given by (4.50), as the first (i.e. dominant) term alone leads to $e^{-(\alpha + \theta_1)y}$ and $e^{-(\alpha - \theta_1)y}$ ($\theta_1 = (\alpha^2 + 1)^{1/2}$). Even though the coefficient of the second of these is small, it is clearly only an approximation to say that the term $2\alpha\,\partial u/\partial y$ leads to a solution that behaves as $e^{-2\alpha y}$. For a numerical solution of (4.56), it is possible to use test functions that are modifications of tensor product trial functions, for example,

$$\psi_{ij} = \varphi_i(x)(\varphi_j(y) + \gamma_j\sigma_j(y)). \tag{4.63}$$

If there had been a $\partial u/\partial x$ term in (4.56) then it would be possible also to modify

$\varphi_i(x)$, using the test functions of the form

$$\psi_{ij} = (\varphi_i(x) + \gamma_x\sigma(x))(\varphi_j(y) + \gamma_y\sigma(y)).$$

Note that

$$\sum_e \int_e \frac{d\varphi_l^e}{dy}\frac{d\sigma_j^e}{dy} = 0 \quad \text{(for all } l \text{ and } j\text{)}$$

but

$$\sum_e \int_e \varphi_l^e\sigma_j^e \neq 0.$$

Thus it follows that in the numerical solution of (4.56), the term corresponding to $\partial^2 u/\partial y^2$ is modified. This leads to the spurious numerical phenomenon called 'cross-wind diffusion' (Hughes and Brooks, 1979). The analysis of Petrov–Galerkin (or upwinding) methods is the subject of much current research and not a little current controversy (Hughes, 1979).

Returning to the numerical solution of (4.56), if the 'optimum' value (see Table 4.11) is selected, i.e.

$$\gamma = \coth(h\alpha) - \frac{1}{h\alpha}$$

with the piecewise constant form of perturbation illustrated in Figure 4.16(c) then the results are again illustrated in Figure 4.15.

Because of the cross-wind dissipation the numerical values usually tend to be below the analytic values although this does not show up in this particular example. In addition for $\alpha = 20$, although the nodal values are reasonably accurate, the accuracy within the elements containing the boundary layer is

Table 4.10 Approximation on graded mesh illustrated in Figure 4.17

Point	$\alpha = 5$				$\alpha = 20$			
$y = -\frac{1}{2}\pi + kh$	F.E.	$(T = 15)$				Uniform Petrov–	Graded Galerkin	
k	Graded grid (a)	Uniform	Exact		k	Galerkin	grid (a)	grid (b)	Exact
$\frac{1}{2}$	1.200		1.187		$\frac{1}{8}$			1.523	1.446
1	1.648	1.882	1.638		$\frac{1}{4}$		2.292	2.044	1.989
$\frac{3}{2}$	1.822		1.817		$\frac{1}{2}$		2.315	2.284	2.270
2	1.897	1.930	1.894		$\frac{3}{4}$		2.318	2.314	2.311
3	1.966	1.966	1.962		1	2.325	2.320	2.320	2.319
					2	2.335	2.330	2.330	2.330

Table 4.11 Optimum values of
$\gamma = \coth(h\alpha) - 1/h\alpha$

α	$h = \pi/8$	$h = \pi/16$
5	0.531	0.308
20		0.7461

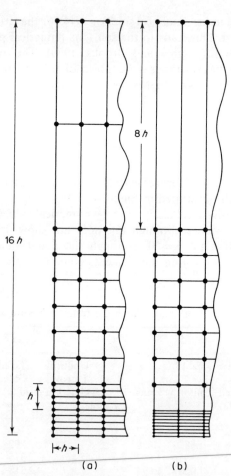

Figure 4.17 $h = \pi/16$. Constant grid spacing in x-direction; graded grid spacing in y-direction

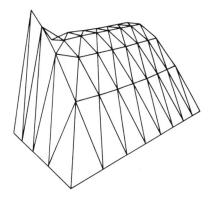

Figure 4.18

generally very poor. It is not possible to obtain an accurate estimate of the size of the boundary layer with such a method.

As there is no $\partial u/\partial x$ term in (4.56), this example corresponds to flow parallel to the grid lines in the y-direction. In general, when the full convective term

$$q_x\frac{\partial u}{\partial x} + q_y\frac{\partial u}{\partial y}$$

is present, it may not be possible to align the grid with the direction of flow and the accuracy of the results can be reduced considerably (Griffiths and Mitchell, 1979).

The alternative frequently suggested is that the next step after the $T = 5$ uniform mesh should be a graded mesh with the same number of nodal parameters. The oscillation is used as a signal that the whole boundary layer lies within the first row of elements. Numerical experiments carried out with the graded grids illustrated in Figure 4.17 provide the results given in Table 4.10. It

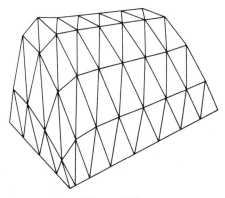

Figure 4.19

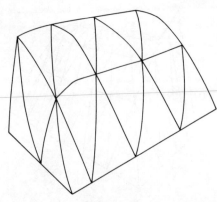

Figure 4.20

can be seen that in all cases the accuracy achieved on either graded grid is better than any results achieved on a uniform grid with the same number of mesh points and the same bilinear basis functions. The boundary conditions are interpolated in the resuts given in Table 4.10 and also in Figure 4.15. The meshes illustrated in Figure 4.17 give the subdivision in the *y*-direction. A uniform spacing $h = \pi/16$ is assumed in the *x*-direction.

Alternative coarse grid solutions are illustrated in Figures 4.18, 4.19, and 4.20. Figure 4.18 shows a perspective surface view of the piecewise linear approximation

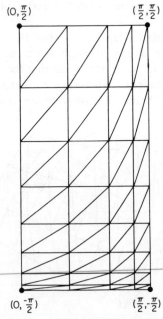

Figure 4.21

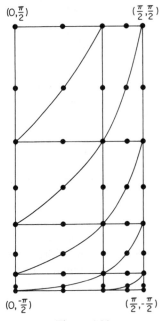

Figure 4.22

on a uniform grid with $T = 7$ and $\alpha = 4$. The smoothly graded grid illustrated in Figure 4.21 is used to generate the solution illustrated in Figure 4.19. In this latter case it can be seen that the spurious wiggles have been almost completely eliminated.

Piecewise isoparametric quadratics with the same nodal positions produce very similar results. The approximation with the graded mesh (Figure 4.22) is given in Figure 4.20: the nodal values are almost identical to those in Figure 4.19. Each of these approximations has 45 nodes and the accuracy obtained is much better than a nonconforming approximation with over 60 nodes either with graded or uniform grids.

All these results were obtained using a FEM teaching package written by one of the authors (R.W.) and which is available in BASIC for a BBC microcomputer.

Exercise 4.8 (a) Verify that the piecewise quadratic and piecewise constant modifications to the test functions lead to the nodal parameters U_i of a piecewise linear approximation to (4.59) defined by

$$\left(\left(1 - \frac{kh}{2}\right) + \gamma_i \frac{kh}{2}\right)U_{i-1} - 2\left(1 + \gamma_i \frac{kh}{2}\right)U_i + \left(\left(1 + \frac{kh}{2}\right) + \gamma_i \frac{kh}{2}\right)U_{i+1} = 0$$

whereas the piecewise linear modification leads to

$$\left(\left(1 - \frac{kh}{2}\right) - \gamma_i\right)U_{i-1} - 2U_i + \left(\left(1 + \frac{kh}{2}\right) + \gamma_i\right)U_{i+1} = 0.$$

(b) Hence show that in either case the solution is

$$U_i = A + B\left(\frac{1-\theta}{1+\theta}\right)^i$$

for some constants A, B, and θ.

(c) Assuming that $U_0 = 1$ and $U_1 = e^{-kh}$ verify (4.60) and (4.62).

Exercise 4.9 Are the three forms of modification equivalent if (4.59) is replaced by

$$\frac{d^2u}{dx^2} + k\frac{du}{dx} + gu = 0?$$

Exercise 4.10 Verify that it follows for (4.60) that $\gamma_i > 0$ if $k > 0$ and $\gamma_i < 0$ if $k < 0$.

4.3.2 Nonphysical models

It has been pointed out many times that the boundary layer phenomenon which caused so much trouble in the preceding examples may not exist in the physical systems being modelled. The boundary layer is often a direct result of the imposition of an incorrect boundary condition at the outflow boundary, i.e.

$$u = 0 \quad y = -\tfrac{1}{2}\pi \quad |x| \leqslant \tfrac{1}{2}\pi$$

in the preceding examples. If this boundary condition is replaced by

$$\frac{\partial u}{\partial y} = 0 \quad y = -\tfrac{1}{2}\pi \quad |x| \leqslant \tfrac{1}{2}\pi$$

then in the solution (4.50), the coefficients $\sinh^{-1}(\tfrac{1}{2}\theta k\pi)$ and $\cosh^{-1}(\tfrac{1}{2}\theta k\pi)$ are replaced by

$$A = \frac{\alpha \cosh\left(\tfrac{1}{2}\theta_k\pi\right) + \theta_k \sinh\left(\tfrac{1}{2}\theta_k\pi\right)}{\alpha \sinh\left(\theta_k\pi\right) + \theta_k \cosh\left(\theta_k\pi\right)}$$

and

$$B = \frac{\alpha \sinh\left(\tfrac{1}{2}\theta_k\pi\right) + \theta_k \cosh\left(\tfrac{1}{2}\theta_k\pi\right)}{\alpha \sinh\left(\theta_k\pi\right) + \theta_k \cosh\left(\theta_k\pi\right)}$$

respectively. This solution has no boundary layer and the solution can be approximated successfully for all values of α.

For example, Figure 4.23 provides a comparison with the solution of the earlier boundary layer example displayed in Figure 4.15. If the uniform mesh is used, both the numerical and the analytic solutions with the derivative boundary condition are indistinguishable from the analytic boundary layer solution outside the boundary layer. In addition, the numerical solution approximates the boundary value very accurately, $u(0, -\tfrac{1}{2}\pi) = 2.309$. With the derivative boundary condition, the larger the value of α, the easier the problem is to solve.

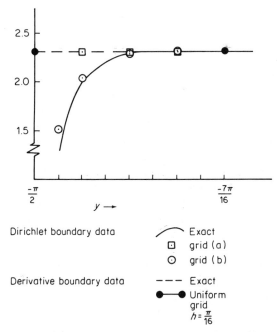

Figure 4.23 Graded mesh Galerkin approximation

4.3.3 Method of collocation

The method of collocation is similar in many respects to Galerkin's method. It involves selecting the coefficients $\alpha_i (i = 1, \ldots, N)$ in the approximation

$$U = \sum_{i=1}^{N} \alpha_i \varphi_i$$

such that the differential equation is satisfied exactly at certain specified points. It has been shown (de Boor and Swartz, 1973; Lucas and Reddien, 1972; also Ahlberg and Ito, 1975) that for ordinary differential equations if the collocation points are chosen correctly, then the method is similar in accuracy to Galerkin's method with the same set of basis functions $\varphi_i (i = 1, \ldots, N)$. If, for example, the basis functions are Hermite piecewise polynomials of degree $2r - 1$, then the collocation points in each subinterval $[x_i, x_{i+1}]$ are taken as the zeros of the Legendre polynomial

$$P_r\left(\frac{2x - x_{i+1} - x_i}{x_{i+1} - x_i}\right).$$

The advantages of the method of collocation are:

(i) There are no inner products to integrate as in Galerkin and Ritz approximations.
(ii) The resultant algebraic equations have fewer terms than the corresponding equations for Galerkin approximations.

The main disadvantage of collocation is that it is necessary to use basis functions of degree (at least) $2k$ for a differential equation of order $2k$.

Methods using collocation have also been devised (Douglas and Dupont, 1973) for evolutionary problems.

As an alternative to the method of collocation, a hybrid collocation–Galerkin method has been proposed by Diaz (1977, 1979) and by Wheeler (1977, 1978). Collocation requires basis functions that have a high degree of continuity, at least C^1, otherwise it is not possible to generate a well posed collocation problem. The basic deficiency of C^0 functions can be illustrated using a 1-dimensional piecewise quadratic with three nodes per element. The nodes of an individual element are the two endpoints plus the midpoint and, globally, if there are N elements there will be $2N + 1$ degrees of freedom in a continuous approximation. Thus it is necessary, in general, to have two collocation points per element to specify all the nodal values. Let us assume that the equation to be solved is

$$u''(x) = f(x)$$

and that in the element $e = [x_i, x_{i+1}]$ the approximate solution is written as

$$U^e(x) = a_i + b_i x + c_i x^2$$

If the collocation points in e are ξ_i and η_i then the two collocation equations are

$$2c_i = f(\xi_i)$$

and

$$2c_i = f(\eta_i).$$

These are clearly inconsistent unless $f(\eta_i) = f(\xi_i)$ in which case the collocation equations do not determine the solution uniquely. Similarly Lagrange cubic elements with two interior nodes imply three collocation points but only two degrees of freedom in the second derivative which is linear. The Hermite cubic has all the nodes at the ends of the elements and as the higher global continuity reduces the total number of degrees of freedom in the approximation to $2N + 2$ (N elements), only two collocation points per element are necessary. Thus Hermite cubics lead to a feasible collocation problem; Lagrange elements, in general, do not. Splines have even higher global continuity and hence still fewer degrees of freedom and so collocation at the nodes leads to a well posed problem with cubic splines. Some of the more interesting properties of spline collocation are outlined in a later subsection.

4.3.4 Collocation–Galerkin method

The basic strategy of collocation–Galerkin is to use piecewise Lagrange elements, collocate at as many points as possible, and then add sufficient (Petrov–) Galerkin equations to provide a well posed problem. Thus, in one dimension, we have collocation equations corresponding to the interior nodes and Galerkin equations corresponding to the nodes at the ends of the elements. As with pure collocation, selecting the collocation points to be coincident with the nodes does not necessarily lead to the optimum method. Diaz (1977, 1979) has shown that the collocation points should be at the zeros of the Jacobi polynomials

$$J_{r-1}\left(\frac{x - x_i}{x_{i+1} - x_i}\right).$$

With $r = 2$, piecewise quadratics, the single collocation point coincides with the single interior node; but for $r = 3$, cubics, they are distinct.

The particular choice of the remaining Petrov–Galerkin equations can be explained in a simple if nonrigorous manner. The interior nodal values are determined by the collocation equations, and hence it is not necessary to provide a full set of Galerkin equations. Without the need for functions corresponding to the interior nodes, it is possible to make use of simple piecewise linear test functions corresponding to the vertex nodes only. Thus given an ordinary differential equation

$$Au = f \tag{4.64}$$

a piecewise quadratic collocation–Galerkin approximation

$$U(x) = \sum_{i=0}^{N} U_i \psi_i(x) + \sum_{j=0}^{N-1} U_{j+1/2} \chi_{j+1/2}(x)$$

is defined by

$$AU(x_{j+1/2}) = f(x_{j+1/2}) \quad j = 1, \ldots, N$$

and

$$a(U, \varphi_i) = (f, \varphi_i) \quad i = 1, \ldots, N-1$$

where it is assumed that

$$a(u, v) = (f, v)$$

is the weak form of (4.64). The shape functions φ_i, ψ_i, and $\chi_{j+1/2}$ are the simple functions shown in Figures 1.3 and 1.4. The nodal values U_0 and U_N are determined by the boundary data or by additional Galerkin equations depending on the form of the boundary conditions (i.e. Dirichlet or Neumann).

In two dimensions, the procedure becomes more complicated. Consider, for example, the 9-node biquadratic element. It has one interior node at which collocation is possible and four corner nodes at which Petrov–Galerkin equations can be defined using piecewise bilinear test functions. At the midside nodes however, it is necessary to adopt a more novel approach. An equation is constructed involving collocation in the tangential direction and integration in the normal direction across two adjacent elements.

In order to explain the method in detail it is necessary to assume that the rectangular region $R = [x_0, x_N] \times [y_0, y_M]$ is partitioned into elements $E_{ij} = [x_i, x_{i+1}] \times [y_j, y_{j+1}]$, $i = 0,\ldots, N-1$; $j = 0,\ldots, M-1$ where the nodes of E_{ij} are identified as $(x_k, y_l) \begin{cases} k = i, i+\frac{1}{2}, i+1 \\ l = j, j+\frac{1}{2}, j+1 \end{cases}$.

As a particular illustration we consider the collocation–Galerkin solution of

$$\nabla^2 u + f = 0 \quad \text{in } R$$

i.e.

$$\frac{\partial^2 u}{\partial x^2} + \frac{\partial^2 u}{\partial y^2} + f(x, y) = 0 \quad \text{in } R$$

with the weak form

$$a(u, v) = (f, v).$$

where

$$a(u, v) = \iint_R \left\{ \frac{\partial u}{\partial x}\frac{\partial v}{\partial x} + \frac{\partial u}{\partial y}\frac{\partial v}{\partial y} \right\} dx\, dy.$$

The biquadratic collocation–Galerkin approximation can be written in terms of tensor products of $\psi_i(x)$ and $\chi_j(x)$ with $\psi_k(y)$ and $\chi_l(y)$. The equations defining the nodal parameters are of four types:

(a) Pure collocation of the interior nodes,

$$\nabla^2 U(x_{i+1/2}, y_{j+1/2}) + f(x_{i+1/2}, y_{j+1/2}) = 0.$$

(b) Petrov–Galerkin equations using bilinear test functions $\varphi_{ij}(x, y) = \varphi_i(x)\varphi_j(y)$ at the corner nodes,

$$a(U, \varphi_{ij}) = (f, \varphi_{ij}).$$

(c) Collocation in x with 'Galerkin' in y at the side nodes on the horizontal edges (i.e. parallel to the x-axis; see Figure 4.24)

$$\int \frac{\partial^2}{\partial x^2} U(x_{i+1/2}, \eta)\varphi_j(\eta)\, d\eta - \int \frac{\partial}{\partial y} U(x_{i+1/2}, \eta)\varphi_j'(\eta)\, d\eta$$

$$+ \int f(x_{i+1/2}, \eta)\varphi_j(\eta)\, d\eta = 0.$$

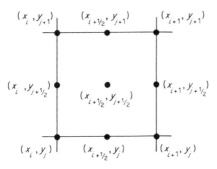

Figure 4.24 Element E_{ij}

Clearly for $0 < j < M$ the range of integration is $[y_{j-1}, y_{j+1}]$, $[y_0, y_1]$ if $j = 0$ and $[y_{M-1}, y_M]$ if $j = M$.

(d) Collocation in y with 'Galerkin' in x corresponding to the nodes on the vertical sides (i.e. parallel to the y axis),

$$- \int \frac{\partial}{\partial x} U(\xi, y_{j+1/2}) \varphi_i(\xi) \, \mathrm{d}\xi + \int \frac{\partial^2}{\partial y^2} U(\xi, y_{j+1/2}) \varphi_i(\xi) \, \mathrm{d}\xi$$

$$+ \int f(\xi, y_{j+1/2}) \varphi_i(\xi) \, \mathrm{d}\xi = 0.$$

It is possible to define collocation Galerkin methods for any tensor product type of trial space; full details can be found in Diaz (1977, 1979). As the pure collocation is associated with the interior node it is possible to adapt the method to the 8-node element by the simple expedient of deleting the pure collocation equations.

4.3.5 Spline collocation

Consider the one-dimensional problem of approximating the solution of the differential equation

$$u''(x) = u(x) \tag{4.65}$$

as a linear combination of B-splines of degree q. Thus

$$U(x) = \sum_i \varphi_i(x) c_i \tag{4.66}$$

where

$$\varphi_i(x) = B_q\left(\frac{x}{h} - i\right).$$

If the unknown coefficients c_i are determined by the Galerkin method it follows that

$$\sum_i c_i \left\{ \int \varphi_i'(x)\varphi_j'(x)\,dx + \int \varphi_i(x)\varphi_i(x)\,dx = 0 \right\} \quad \forall j. \tag{4.67}$$

If $X = x/h - i$ and $i - j = Y$, then

$$\int \varphi_i(x)\varphi_j(x)\,dx = \int B_q\left(\frac{x}{h}-i\right)B_q\left(\frac{x}{h}-j\right)dx$$

$$= h \int B_q(X)B_q(X-Y)\,dX$$

$$= hB_{2q+1}(Y). \tag{4.68}$$

Similarly

$$\int \varphi_i'(x)\varphi_j'(x)\,dx = \int B_q'\left(\frac{x}{h}-i\right)B_q'\left(\frac{x}{h}-j\right)dx$$

$$= \frac{1}{h}\int B_q'(X)B_q'(X-Y)\,dx$$

$$= -\frac{1}{h}\int B_q(X)B_q''(X-Y)\,dx$$

$$= -\frac{1}{h}B_{2q+1}''(Y). \tag{4.69}$$

If we define

$$\psi_k(x) = B_{2q+1}\left(\frac{x}{h}-k\right)$$

then as *one* possible substitution for Y is

$$Y = -\left(\frac{jh}{h}-i\right)$$

and as

$$B(X) = B(-X)$$

it follows that

$$B_{2q+1}(Y) = \psi_i(jh) \tag{4.70}$$

and that

$$B_{2q+1}''(Y) = h^2\psi_i(jh). \tag{4.71}$$

Thus using (4.67)–(4.70) it is possible to write the Galerkin equations (4.66) as collocation equations for an approximation

$$V(x) = \sum_i \psi_i(x)c_i. \tag{4.72}$$

That is, (4.66) is equivalent to

$$\sum_i c_i\{\psi_i''(jh) - \psi_i(jh)\} = 0 \quad \forall j \tag{4.73}$$

i.e. collocation at the node $x_j = jh$.

For two-dimensional problems this means that the Galerkin solution of Laplace's equation in terms of piecewise bilinears ($q = 1$) leads to the same equations as a collocation solution in terms of piecewise bicubics ($2q + 1 = 3$).

In the case of Poisson's equations with an inhomogeneous term, i.e.

$$\frac{\partial^2 u}{\partial x^2} + \frac{\partial^2 u}{\partial y^2} = f(x, y)$$

the equivalence is only true if the function $f(x, y)$ can be written as the tensor product of splines of degree at most q.

The equations (4.67) and (4.73) are in terms of the coefficients c_i in the respective spline approximations (4.66) and (4.72) and not in terms of the nodal values $U_i \equiv U(ih)$ and $V_i \equiv V(ih)$. But in certain circumstances, with homogeneous boundary conditions the nodal values satisfy the same equations as the parameters c_i (see Vichnevetsky and Bowles 1982). The proof relies on the fact that the matrices

$$A = \{a_{ij}\} = \{\varphi_i(jh)\} \tag{4.74}$$

and

$$B = \{b_{jk}\} = \{\varphi_j''(kh)\} \tag{4.75}$$

commute (see Exercise 4.12).

Exercise 4.11 Verify that

$$B_{2q+1}(Y) = \psi_j(ih)$$
$$B_{2q+1}'(Y) = h\psi_j'(ih)$$

and

$$B_{2q+1}''(Y) = h^2\psi_j''(ih)$$

and show that

$$\int \phi_i(x)\phi_j'(x)\,dx = -B_{2q+1}'(Y).$$

Exercise 4.12 (a) Verify numerically that for cubic splines

$$\sum_j \psi_i''(jh)\psi_j(kh) = \sum_j \psi_i(jh)\psi_j''(kh).$$

(b) Assuming that the matrices (4.74) and (4.75) commute, multiply (4.73) by $\psi(kh)$ and sum over j to prove that

$$\sum_j V_j\{\psi_j''(kh) - \psi_j(kh)\} = 0 \quad \forall k$$

i.e. the $\{V_j\}$ and $\{c_i\}$ satisfy equivalent systems.

4.3.6 H^{-1} and H^1 Galerkin methods

If the test and/or trial functions are smoother than the simple piecewise linear or bilinear used in the preceding examples, it is possible to use a variety of methods that require higher order continuity. In Section 4.2 it was shown that since $a(u, v)$ and (Au, v) only differ by boundary terms, the Galerkin solution and the classical solution are the same. In a similar manner it follows that since (Au, v) and (u, A^*v) only differ by boundary terms, the H^{-1} Galerkin equations

$$(u, A^*v) = (f, v) \qquad \forall v \in \mathscr{H} \tag{4.76}$$

have the same solution as the original form.

Clearly v must be sufficiently differentiable for (4.76) to exist. In the example earlier ((4.28) *et seq.*)

$$A = A^* = -\frac{\partial}{\partial x}\left(d_1 \frac{\partial}{\partial x}\right) - \frac{\partial}{\partial y}\left(d_2 \frac{\partial}{\partial y}\right)$$

i.e. the operator is self adjoint. It then follows that

$$(Au, v) - (u, A^*v) = -\int_{\partial R}\left(n_1 d_1 \frac{\partial u}{\partial x} + n_2 d_2 \frac{\partial u}{\partial y}\right)v \, ds \tag{4.77}$$

$$+ \int_{\partial R}\left(n_1 d_1 \frac{\partial v}{\partial x} + n_2 d_2 \frac{\partial v}{\partial y}\right)u \, ds.$$

It is clear from the terms on the right-hand side of (4.77) which boundary conditions should be imposed.

Assume that the given boundary condition is

$$u = 0. \tag{4.78}$$

It follows that condition

$$v = 0$$

must be *imposed* on the test space $\mathscr{H}^{(1)} \cap \mathscr{H}^{(2)}$. In such a case the condition (4.78) is the natural boundary condition resulting from the H^{-1} formulation (4.76).

The H^1 Galerkin method requires that the $\mathscr{L}_2$ inner product is replaced by the $\mathscr{H}^{(1)}$ inner product in the equation

$$(Au, v) = (f, v).$$

Such results as are available (see Mitchell and Griffiths, 1977) for the H^1 and H^{-1} methods are not particularly encouraging. Theoretical results for H^{-1} and H^1 methods can be found in Rachford and Wheeler (1974) and Douglas *et al.* (1974), respectively.

Method of least squares

An alternative use of smooth test functions is to define an approximation U as the solution of

$$\underset{U \in K_N}{\text{minimum}} \| AU - Au_0 \|^2$$

or

$$\underset{U \in K_N}{\text{minimum}} \| AU - f \|^2,$$

where $\| \cdot \|$ is the $\mathcal{L}^2$ norm. This is the basis of the method of least squares (Bramble and Schatz, 1970, 1971). If we apply the calculus of variations to determine the conditions necessary for a stationary value of such a functional, we obtain an Euler–Lagrange equation involving the operator A^*A, that is, an equation of higher order than the one given. For this reason it follows that we must take care to ensure that the solutions are the same, particularly when inhomogeneous Dirichlet boundary conditions are specified; thus unless we impose additional conditions on the approximating functions, we must modify the functional (Section 4.2). If we are given

$$u = g$$

on ∂R, a least-squares approximation could minimize

$$\int\int_R (AU - f)^2 \, dx + c \int_{\partial R} (U - g)^2 \, ds$$

for a suitable choice of c (cf. the Galerkin method with a penalty functional, Section 4.2.6).

4.4 Nonlinear Equations and Systems

The Galerkin method can be applied to nonlinear problems, but it is only in special cases that it is possible to derive a weak form that has reduced continuity requirements. One example for which there is such a reduction is the nonlinear equation

$$\frac{\partial}{\partial x}\left(p(u)\frac{\partial u}{\partial x} \right) + \frac{\partial}{\partial y}\left(q(u)\frac{\partial u}{\partial y} \right) + f(x, y) = 0. \tag{4.79}$$

Define

$$a(u, v; w) = \int\int_R \left\{ p(w)\frac{\partial u}{\partial x}\frac{\partial v}{\partial x} + q(w)\frac{\partial u}{\partial y}\frac{\partial v}{\partial y} \right\} dx \, dy \tag{4.80}$$

then the weak form of (4.79) can be written as

$$a(u, v; u) - (f, v) = 0 \quad \forall v \in \mathcal{H}. \tag{4.81}$$

This leads to Galerkin equations for the approximate solution in the form

$$a(U, \varphi_i; U) = \int\int_R \left\{ p(U)\left(\frac{\partial U}{\partial x}\right)\left(\frac{\partial \varphi_i}{\partial x}\right) + q(U)\left(\frac{\partial U}{\partial y}\right)\left(\frac{\partial \varphi_i}{\partial y}\right) \right\} dx \, dy$$

$$= (f, \varphi_i) \quad (i = 1, \ldots, N) \tag{4.82}$$

If $p = q$ then equation (4.79) can be viewed as a steady state diffusion equation. In such an interpretation $p\,(=q)$ is the diffusivity, thus $p = p(u)$ implies that the diffusivity depends on the concentration. Say, the wetter the soil the more easily water diffuses through it or, if u is the temperature, the thermal conductivity is temperature dependent.

In terms of the nonlinear stiffness matrix

$$G(\mathbf{U}) = \left\{ \int\int_R \left(p(U)\frac{\partial\varphi_j}{\partial x}\frac{\partial\varphi_i}{\partial x} + q(U)\frac{\partial\varphi_j}{\partial y}\frac{\partial\varphi_i}{\partial y} \right)dx\,dy \right\}$$

the Galerkin equations (4.82) can be written as

$$G(\mathbf{U})\mathbf{U} = \mathbf{b} \qquad (4.83)$$

where

$$U = \sum_j U_j\varphi_j(\mathbf{x}) \qquad \mathbf{U} = [U_1,\ldots,U_N]^T$$

and

$$b_i = (f, \varphi_i).$$

It is necessary to devise a method of solution for the nonlinear equations (4.82). In the preceding linear examples it is assumed that a system of linear Galerkin equations is solved by a veriant of Gauss elimination; such solution techniques are discussed in Section 4.5. A system of nonlinear equations is usually linearized in some way in order to generate a sequence of iterates that converge to the solution of the nonlinear problem. For example, we could generate a sequence of solutions

$$U^{(k)}(x, y) = \sum_i U_i^{(k)}\varphi_i(x, y)$$

$k = 1, 2,\ldots$ such that (given $U^{(k-1)}$) $U^{(k)}$ is the solution of

$$\int\int \left\{ p(U^{(k-1)})\frac{\partial U^{(k)}}{\partial x}\frac{\partial\varphi_i}{\partial x} + q(U^{(k-1)})\frac{\partial U^{(k)}}{\partial y}\frac{\partial\varphi_i}{\partial y} \right\}dx\,dy = (f, \varphi_i) \qquad (4.84)$$

i.e.

$$a(U^{(k)}, \varphi_i; U^{(k-1)})(f, \varphi_i) \quad (i = 1,\ldots, N). \qquad (4.85)$$

Frequently the nonlinear terms themselves are simplified so the $p(U)$ is interpolated using the basis functions. Thus if

$$\tilde{p}_i = p(U(x_i, y_i)) \quad i = 1,\ldots, N$$

the function $p(u)$ is replaced by the interpolant

$$p_h = \sum_i \tilde{p}_i\varphi_i(x, y).$$

The replacement of a nonlinear term by a linear combination of the basis functions is an example of *product approximation* and will be used in Chapters 5 and 7.

If the function q is similarly replaced by an interpolant q_h, the Galerkin equations then become

$$a_h(U, \varphi_i; U) = \int\int \left\{ p_h \frac{\partial U}{\partial x} \frac{\partial \varphi_i}{\partial x} + q_h \frac{\partial U}{\partial y} \frac{\partial \varphi_i}{\partial y} \right\} = (f, \varphi_i) \quad (i = 1, \ldots, N) \quad (4.86)$$

and the iteration can be written as

$$\int\int \left\{ p_h^{(k-1)} \frac{\partial U^{(k)}}{\partial x} \frac{\partial \varphi_i}{\partial x} + q_h^{(k-1)} \frac{\partial U^{(k)}}{\partial y} \frac{\partial \varphi_i}{\partial y} \right\} = (f, \varphi_i) \quad (i = 1, \ldots, N) \quad (4.87)$$

i.e.

$$a_h(U^{(k)}, \varphi_i; U^{(k-1)}) = (f, \varphi_i).$$

One considerable computational advantage to be gained from interpolating the nonlinearities is that the element matrices are much easier to construct. The approximate element matrix $A_h^{(k)e}$ can split up as a linear combination of matrices B_i, C_i for each node in element e, where

$$(B_i^e)_{jl} = \int\int_e \varphi_i \frac{\partial \varphi_j}{\partial x} \frac{\partial \varphi_l}{\partial x} \tag{4.88a}$$

and

$$(C_i^e)_{jl} = \int\int_e \varphi_i \frac{\partial \varphi_j}{\partial y} \frac{\partial \varphi_l}{\partial y}. \tag{4.88b}$$

These matrices remain constant and do not vary as the iteration proceeds.

In terms of the nonlinear system (4.83), the simple iteration scheme can be written as

$$G(U^{(k-1)})U^{(k)} = \mathbf{b}.$$

If we wish to solve the quasi-linear equation

$$\frac{\partial^2 u}{\partial x^2} + \frac{\partial^2 u}{\partial y^2} = f(u)$$

then product approximation can be used to replace the nonlinear term $f(U)$ where

$$U = \sum_j U_j \varphi_j(x)$$

by

$$f_h(U) = \sum_j f(U_j)\varphi_j(x)$$

and thus the Galerkin equations become

$$a(U, \varphi_i) + (f_h(U), \varphi_i) = 0 \quad (i = 1, \dots, N).$$

These can be written as the linearized algebraic equation

$$GU + Bf = 0$$

where

$$G = \{a(\varphi_j, \varphi_i)\}$$

is the stiffness matrix,

$$B = \{(\varphi_j, \varphi_i)\}$$

is the so-called *mass matrix* and

$$\mathbf{f} = [f(U_1), \dots, f(U_N)]^T.$$

Simple iteration for such problems can then be written as

$$GU^{(k+1)} + Bf^{(k)} = 0$$

or alternatively as

$$a(U^{(k+1)}, \varphi_i) + (f_h(U^{(k)}), \varphi_i) = 0.$$

If the simple iteration (4.85) or (4.87) fails to converge, it is possible to use a Newton method (Douglas and Dupont, 1975). In order to derive this method, it is necessary to introduce a little additional material.

If (4.81) is written as

$$G(u) = 0 \qquad \forall v \in \mathscr{H}$$

then the Newton iteration provides a sequence of iterates $\{u^{(k)}\}$ such that

$$G'(u^{(k)})(u^{(k+1)} - u^{(k)}) = -G(u^{(k)}) \quad \forall v \in \mathscr{H}. \tag{4.89}$$

The linear operator G' is the *Gateaux differential (operator)* defined by

$$G'(u)w = \lim_{\varepsilon \to 0} \frac{1}{\varepsilon}[G(u + \varepsilon w) - G(u)]. \tag{4.90}$$

If $G(u)$ is defined by (4.80) and (4.81) then direct evaluation of (4.90) leads to

$$G'(u)w = a'(w, v; u)$$

$$= \iint \left\{ p(u)\frac{\partial w}{\partial x}\frac{\partial v}{\partial x} + p_u(u)\frac{\partial u}{\partial x}w\frac{\partial v}{\partial x} + q(u)\frac{\partial w}{\partial y}\frac{\partial v}{\partial y} + q_u(u)\frac{\partial u}{\partial y}w\frac{\partial v}{\partial y} \right\} dx\,dy. \tag{4.91}$$

Thus in this notation the Newton iteration formula is

$$a'(u^{(k+1)} - u^{(k)}, v; u^{(k)}) = (f, v) - a(u^{(k)}, v; u^{(k)}) \quad \forall v. \tag{4.92}$$

The sequence of iterates $U^{(k)}$ is therefore generated by the equations

$$a'(U^{(k+1)} - U^{(k)}, \varphi_i; U^{(k)}) = (f, \varphi_i) - a(U^{(k)}, \varphi_i; U^{(k)}) \quad (i = 1, \dots, N).$$

These are linear in the corrections $U_j^{(k+1)} - U_j^{(k)}$ ($j = 1, \ldots, N$) but again the computation would be simplified if p and q were replaced by p_h and q_h respectively and if in addition

$$p_u(U)\frac{\partial U}{\partial x}$$

is replaced by

$$p_h'(U) = \sum_i \tilde{p}_i \varphi_i(x, y)$$

where

$$\tilde{p}_i = p_u(U(x_i, y_i))\frac{\partial U(x_i, y_i)}{\partial x}.$$

Similar product approximations can be defined for the term $q_u(U)\partial U/\partial x$. The sequence of numerical approximations $\{U_h^{(k)}\}$ is then defined by

$$a_h'(U_h^{(k+1)} - U_h^{(k)}, \varphi_i; U_h^{(k)}) = (f, \varphi_i) - a_h(U_h^{(k)}, \varphi_i; U_h^{(k)})$$

where in this case

$$a_h'(W, V; U) = a_h(W, V; U) + \int\int_R \left\{ p_h'(U)W\frac{\partial V}{\partial x} + q_h'(U)W\frac{\partial V}{\partial y} \right\} \qquad (4.93)$$

and

$$a_h(W, V; U) = \int\int \left\{ p_h(U)\frac{\partial W}{\partial x}\frac{\partial V}{\partial x} + q_h(U)\frac{\partial W}{\partial y}\frac{\partial V}{\partial y} \right\}. \qquad (4.94)$$

Thus we solve the system

$$G_h'^{(k)}(U_h^{(k+1)} - U_h^{(k)}) = f_h^{(k)} \qquad (4.95)$$

where the *element matrix* $G_h'^e$ is a linear combination of B_i^e, C_i^e,

$$(D_i^e)_{jl} = \int\int_e \varphi_i \varphi_l \frac{\partial \varphi_j}{\partial x}$$

and

$$(E_i^e)_{jl} = \int\int_e \varphi_i \varphi_l \frac{\partial \varphi_j}{\partial y}.$$

Alternatively the system (4.83) can be viewed as a nonlinear algebraic system solved by Newton or quasi-Newton methods (Matthies and Strang, 1979).

An adaptive method in which the sequence of approximations $\{u^{(k)}\}$ in (4.89) is defined using progressively finer grids is known as the *projective Newton method*, and has been studied by Witsch (1978a, b).

4.4.1 Numerical example 4

Table 4.12 gives the results of the simple iteration (4.85) applied to obtain a

Table 4.12 Solution of (4.96) with $n = 1$. Grid as in Figure 4.4
with interpolated boundary data

Point	Solution		
	$T = 7$	$T = 15$	Exact
1	0.387	0.440	0.446
2	0.927	0.983	1.000
3	2.139	2.180	2.193
4	0.256	0.307	0.322
5	0.642	0.692	0.707
6	1.509	1.541	1.551
Iterations	23	18	(simple iteration)
	11	22	(Newton iteration)

piecewise bilinear approximation to the solution of

$$\frac{\partial}{\partial x}\left(u^n \frac{\partial u}{\partial x}\right) + \frac{\partial}{\partial y}\left(u^n \frac{\partial u}{\partial y}\right) = n\,e^{(n+1)y}\cos^{n-1}x \tag{4.96}$$

in

$$R = [-\tfrac{1}{2}\pi, \tfrac{1}{2}\pi] \times [-\tfrac{1}{2}\pi, \tfrac{1}{2}\pi]$$

with $n = 1$ and Dirichlet boundary conditions compatible with the solution

$$u(x, y) = e^y \cos x.$$

The approximation assumes the boundary condition is interpolated by the basis functions. The uniform meshes of earlier examples were used in this calculation. Note that for the case $n = 1$, we have $p = \tilde{p}$ and $q = \tilde{q}$ since the given functions are linear in $\tilde{u}$.

In each case the iteration is stopped when the numerical solutions converge to 3 decimal places. It can be seen from Table 4.12 that once again the error in the nodal parameters decreases as $O(h^2)$.

In both cases the initial approximation was $U^{(0)} = 1$ and so $U^{(1)}$ is the solution of the linear problem

$$\frac{\partial^2 u}{\partial x^2} + \frac{\partial^2 u}{\partial y^2} = n\,e^{(n+1)y}\cos^{n-1}x$$

subject to the same boundary conditions.

As can be observed from Table 4.11, Newton's method will sometimes converge more quickly than simple iteration, but neither method can be guaranteed to work in all cases. Indeed if both methods one applied to (4.96) with $n = 2$, simple iteration diverges so rapidly that arithmetic overflow occurs after 10 iterations, whereas the linearized Newton (i.e. using (4.93)–(4.95)) converges at most but not all nodal values.

4.4.2 Systems of equations

The small displacement theory of elastic solids (plane stress), see Chapter 2, can give rise to a pair of differential equations of the form

$$u_{xx} + \frac{(1-v)}{2} u_{yy} + \frac{(1+v)}{2} v_{xy} = 0$$

and

$$v_{yy} + \frac{(1-v)}{2} v_{xx} + \frac{(1+v)}{2} u_{xy} = 0. \tag{4.97}$$

For such problems in which the differential system can be derived via a variational functional $I(u, v)$ the Ritz method provides an unambiguous method of approximation. Consider approximations

$$U = \sum_{i=1}^{N} U_i \varphi_i$$

and

$$V = \sum_{j=1}^{M} V_j \psi_j$$

and it does not matter whether $N = M$ or whether $\{\varphi_i\} = \{\psi_i\}$.
The variational equations

$$\frac{\partial}{\partial U_i} I(U, V) = 0 \quad (i = 1, \ldots, N)$$

and

$$\frac{\partial}{\partial V_i} I(U, V) = 0 \quad (i = 1, \ldots, M) \tag{4.98}$$

provide the correct number of equations for determining the unknowns $U_i (i = 1, \ldots, N)$ and $V_i (i = 1, \ldots, M)$.

Alternatively, if there is no variational principle available the Galerkin method leads to a number of possible solutions. If the weak form of the equations is

$$a_1(u, v, w_1) = (f_1, w_1) \quad \forall w_1 \in \mathcal{H}_1$$

and

$$a_2(u, v, w_2) = (f_2, w_2) \quad \forall w_2 \in \mathcal{H}_2$$

the approximation problem can be *either*

$$a_1(U, V, \varphi_i) = (f_1, \varphi_i) \quad (i = 1, \ldots, N)$$

and
$$a_2(U, V, \psi_i) = (f_2, \psi_i) \quad (i = 1, \ldots, M)$$
or
$$a_1(U, V, \psi_i) = (f_1, \psi_i) \quad (i = 1, \ldots, M)$$
and
$$a_2(U, V, \varphi_i) = (f_2, \varphi_i) \quad (i = 1, \ldots, N).$$

In most problems it is obvious which formulation should be used; for example with the equilibrium equations (4.97), the first equation specifies equilibrium in the x-direction and u is the displacement in the x-direction. Thus the first formulation is appropriate in this case.

Another problem associated with identifying the appropriate Galerkin equations corresponding to the correct natural boundary conditions is illustrated by the numerical example in the next section. The deformation of elastic solids is one of the areas where the Ritz formulation is preferred to the Galerkin formulation. The reason is simply that the problem arises naturally as a variational principle, viz. minimize the potential energy. As mentioned earlier, a Ritz finite element solution can be derived via an alternative Galerkin formulation; the verification of the equivalence in this example is left as Exercise 4.4.

4.4.3 Numerical example 5: plane stress

As an example, the equations (4.97) are solved in the region R shown in Figure (4.25(a):
$$u = v = 0 \quad \text{on } AB \tag{4.99a}$$
and
$$u = v = 0.3 \quad \text{on } DE$$

with the natural boundary conditions
$$n_1 \sigma_x + n_2 \sigma_{xy} = 0$$
and
$$n_1 \sigma_{xy} + n_2 \sigma_y = 0 \tag{4.99b}$$

on the boundaries BD and EA. The natural boundary conditions represent edges free from surface tractions given that
$$\sigma_x = \left(\frac{\partial u}{\partial x} + v \frac{\partial v}{\partial y} \right) \frac{E}{1 - v^2} \tag{4.99c}$$

$$\sigma_y = \left(\frac{\partial v}{\partial y} + v \frac{\partial u}{\partial x} \right) \frac{E}{1 - v^2}$$

and
$$\sigma_{xy} = \frac{1 - v}{2} \left(\frac{\partial u}{\partial y} + \frac{\partial v}{\partial x} \right) \frac{E}{1 - v^2}.$$

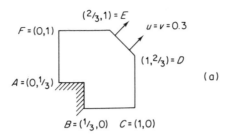

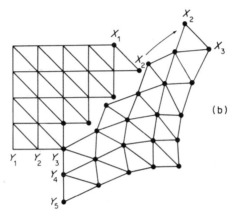

Figure 4.25 Plane-stress linear elastic deformation

As the boundary conditions imply that either both $u(x, y)$ and $v(x, y)$ are specified, or both are unknown, the numerical solution is of the form

$$U = \sum_{j=1}^{N} U_j \varphi_j$$

and

$$V = \sum_{j=1}^{N} V_j \varphi_j.$$

Thus the nodal parameters are in pairs giving the x-displacement and y-displacement of each node.

The Galerkin equations are then

$$\iint \left\{ \frac{\partial U}{\partial x} \frac{\partial \varphi_i}{\partial x} + v \frac{\partial V}{\partial y} \frac{\partial \varphi_i}{\partial x} + \frac{(1-v)}{2} \left(\frac{\partial U}{\partial y} + \frac{\partial V}{\partial x} \right) \frac{\partial \varphi_i}{\partial y} \right\} = 0$$

$$i = 1, \ldots, N \quad (4.100)$$

and

$$\iint \left\{ \frac{\partial V}{\partial y} \frac{\partial \varphi_i}{\partial y} + v \frac{\partial U}{\partial x} \frac{\partial \varphi_i}{\partial y} + \frac{(1-v)}{2} \left(\frac{\partial U}{\partial y} + \frac{\partial V}{\partial x} \right) \frac{\partial \varphi_i}{\partial x} \right\} = 0$$

$$i = 1, \ldots, N. \quad (4.101)$$

This system of equations can be written as

$$\begin{bmatrix} K_{11} & K_{12} \\ K_{21} & K_{22} \end{bmatrix} \begin{bmatrix} \mathbf{U} \\ \mathbf{V} \end{bmatrix} = \begin{bmatrix} \mathbf{f} \\ \mathbf{g} \end{bmatrix} \tag{4.102}$$

where

$$(K_{11})_{ij} = \iint \left\{ \frac{\partial \varphi_j}{\partial x} \frac{\partial \varphi_i}{\partial x} + \frac{(1-v)}{2} \frac{\partial \varphi_j}{\partial y} \frac{\partial \varphi_i}{\partial y} \right\}$$

$$(K_{22})_{ij} = \iint \left\{ \frac{\partial \varphi_j}{\partial y} \frac{\partial \varphi_i}{\partial y} + \frac{(1-v)}{2} \frac{\partial \varphi_j}{\partial x} \frac{\partial \varphi_i}{\partial x} \right\}$$

$$(K_{12})_{ij} = \iint \left\{ v \frac{\partial \varphi_j}{\partial y} \frac{\partial \varphi_i}{\partial x} + \frac{(1-v)}{2} \frac{\partial \varphi_j}{\partial x} \frac{\partial \varphi_i}{\partial y} \right\}$$

and

$$(K_{21})_{ij} = \iint \left\{ v \frac{\partial \varphi_j}{\partial x} \frac{\partial \varphi_i}{\partial y} + \frac{(1-v)}{2} \frac{\partial \varphi_j}{\partial y} \frac{\partial \varphi_i}{\partial x} \right\}$$

i.e.

$$K_{21} = K_{12}^T.$$

The inhomogeneous terms $\mathbf{f}$ and $\mathbf{g}$ only arise from the given boundary displacements since there are no body forces in this example. Thus

$$\mathbf{f} = -(K_{11}\mathbf{U}_\delta + K_{12}\mathbf{V}_\delta)$$

and

$$\mathbf{g} = -(K_{21}\mathbf{U}_\delta + K_{22}\mathbf{V}_\delta)$$

where $\mathbf{U}_\delta$ and $\mathbf{V}_\delta$ are vectors of nodal parameters that interpolate the boundary data. In this example

$$\mathbf{U}_\delta = \begin{cases} 0.3 \text{ for nodes } X_1, X_2, X_3 \\ 0 \text{ for nodes } Y_1, Y_2, Y_3, Y_4, Y_5 \end{cases}$$

and

$$\mathbf{V}_\delta = \begin{cases} 0.3 \text{ for nodes } X_1, X_2, X_3 \\ 0 \text{ for nodes } Y_1, Y_2, Y_3, Y_4, Y_5. \end{cases}$$

If the region R is divided into 60 triangular elements (as in the upper part of Figure 4.25(b)) with piecewise linear $\varphi_i(x, y)$ then with $v = 0.3$ (steel) the finite element approximation is shown in the lower part of Figure 4.25(b). For computational efficiency the unknowns are ordered as

$$(U_1, V_1, \ldots, U_N, V_N)$$

rather than

$$(U_1, \ldots, U_N, V_1, \ldots, V_N)$$

as in (4.101). The former ordering leads to a reduced bandwith in the stiffness matrix.

It is not claimed that the results of Figure 4.25 illustrate a particularly accurate solution. On the contrary, it is widely accepted that both linear triangles and bilinear rectangles have serious limitations in their ability to represent bending modes of deformation. For this reason the 8-node quadrilateral is a very popular element in structural analysis codes.

4.4.4 The Biharmonic equation

As an alternative example, consider the solution of the biharmonic problem

$$\nabla^2(\nabla^2 w) = f \quad \text{in } R$$

subject to

$$w = g_1 \quad \text{on } \partial R$$

and

$$\nabla^2 w = g_2 \quad \text{on } \partial R.$$

The reason for this choice of boundary conditions is that it is then possible to write

$$v = \nabla^2 w$$

and replace the problem by an equivalent system of second-order equations

$$\left. \begin{array}{r} \nabla^2 v = f \\ \nabla^2 w - v = 0 \end{array} \right\} \quad \text{in } R$$

with

$$\left. \begin{array}{r} w = g_1 \\ v = g_2 \end{array} \right\} \quad \text{on } \partial R.$$

Clearly these equations do not have to be solved simultaneously. Assuming approximations of the form

$$W = \sum_{i=1}^{N} W_i \varphi_i$$

and

$$V = \sum_{i=1}^{N} V_i \varphi_i$$

the Galerkin equations are

$$\iint_R \left\{ \frac{\partial V}{\partial x} \frac{\partial \varphi_i}{\partial x} + \frac{\partial V}{\partial y} \frac{\partial \varphi_i}{\partial y} \right\} dx\, dy = - \iint_R f \varphi_i \, dx\, dy \quad i = 1, \ldots, N$$

and

$$\iint_R \left\{ \frac{\partial W}{\partial x} \frac{\partial \varphi_i}{\partial x} + \frac{\partial W}{\partial y} \frac{\partial \varphi_i}{\partial y} \right\} dx\, dy - \iint_R V \varphi_i \, dx\, dy = 0 \quad i = 1, \ldots, N.$$

Thus if

$$(A)_{ij} = \iint_R \left\{ \frac{\partial \varphi_i}{\partial x} \frac{\partial \varphi_j}{\partial x} + \frac{\partial \varphi_i}{\partial y} \frac{\partial \varphi_j}{\partial y} \right\} dx\, dy$$

and

$$(B)_{ij} = \iint_R \varphi_i \varphi_j\, dx\, dy$$

the system equation to be solved can be written as

$$\begin{bmatrix} A & 0 \\ -B & A \end{bmatrix} \begin{bmatrix} \mathbf{V} \\ \mathbf{W} \end{bmatrix} = \begin{bmatrix} \mathbf{f} \\ 0 \end{bmatrix}$$

where $f_i = \iint_R \varphi_i f\, dx\, dy$. As in the preceding numerical example, for computational purposes an efficient ordering of the unknowns would be

$$(V_1, W_1, V_2, W_2, \ldots, V_N, W_N)$$

rather than

$$(V_1, \ldots, V_N, W_1, \ldots, W_N)$$

Exercise 4.13 Derive the Newton method analogous to (4.95) for the mildly nonlinear equation

$$\frac{\partial^2 u}{\partial x^2} + \frac{\partial^2 u}{\partial y^2} = f(u).$$

Exercise 4.14 (a) Determine the (nonphysical) natural boundary conditions corresponding to the Galerkin equations

$$\iint \left\{ \frac{\partial u}{\partial x} \frac{\partial \varphi_i}{\partial x} + \frac{1+v}{2} \frac{\partial v}{\partial y} \frac{\partial \varphi_i}{\partial x} + \frac{1-v}{2} \frac{\partial u}{\partial y} \frac{\partial \varphi_i}{\partial y} \right\} = 0 \quad i = 1, \ldots, N$$

and

$$\iint \left\{ \frac{\partial v}{\partial y} \frac{\partial \varphi_i}{\partial y} + \frac{1+v}{2} \frac{\partial u}{\partial x} \frac{\partial \varphi_i}{\partial y} + \frac{1-v}{2} \frac{\partial v}{\partial x} \frac{\partial \varphi_i}{\partial x} \right\} = 0 \quad i = 1, \ldots, N.$$

What are the significant changes from (4.100)?

(b) Verify that the Ritz formulation (4.98) with $\psi_i = \varphi_i$ leads to the same equations as the 'correct' Galerkin approximation (4.100) when applied to the form of the potential energy given in Chapter 2.

4.5 Computational Considerations

This section is a brief introduction to some of the practical details that have to be resolved in any computer implementation of the mathematical methods of the preceding sections. Any efficient implementation will take full account of the features of the particular hardware to be used (microprocessor, array processor, virtual machine?) but it is possible to make a few general observations. Full details of implementations suitable for microprocessor systems are the subject of a companion volume (Wait, 1985a) which covers all the stages of a complete finite element analysis; from the generation of the mesh, through the evaluation and solution of the algebraic equations, to the graphical output of the results. An associated software package (Wait, 1985b) is also available. There are many other books aimed at improving the coding of finite element routines (Hinton and Owen, 1979; Akin, 1982; Smith, 1982) and a large number of commercially available finite element software packages for solving problems in a variety of applications.

Numerical integration and numerical linear algebra are at the heart of finite element computation and this section is restricted to these two aspects only.

4.5.1 Constructing element matrices

In general the components of the element matrices are computed using numerical quadrature; thus it is necessary to evaluate the integrand of a_{ij}^e (cf. (4.9)) in the reference element, at points determined by the quadrature rule.

Each component of the element matrix is an integral, and so it is necessary to evaluate every integrand at each quadrature point. For example, assume that the 3×3 element matrix

$$G^e = \left\{ \iint_e \left(\frac{\partial \varphi_i^e}{\partial x} \frac{\partial \varphi_j^e}{\partial x} + \frac{\partial \varphi_i^e}{\partial y} \frac{\partial \varphi_j^e}{\partial y} \right) dx \, dy \right\}$$

is to be evaluated using numerical quadrature. A two-point quadrature scheme is defined with weights w_1 and w_2 associated with points (p_1, q_1) and (p_2, q_2) in the reference element $\mathscr{E}$. Then for each quadrature point (p_k, q_k) $(k = 1, 2)$ it is necessary to evaluate the matrix

$$\tilde{G}_k^e = \left\{ \left(\frac{\partial \varphi_i^e}{\partial x} \frac{\partial \varphi_j^e}{\partial x} + \frac{\partial \varphi_i^e}{\partial y} \frac{\partial \varphi_j^e}{\partial y} \right) \frac{\partial(x, y)}{\partial(p, q)} \right\}_{\substack{p = p_k \\ q = q_k}}.$$

The element matrix A^e is then replaced by the quadrature estimate

$$\tilde{G}^e = w_1 \tilde{G}_1^e + w_2 \tilde{G}_2^e.$$

If the reference element is the unit square, there are a number of possible quadrature schemes:

(1) *Newton–Cotes formulae* will often evaluate the integrand at the nodes.
(2) *Gauss formulae* will provide higher order for a given number of evaluations:

$$\int_0^1 \int_0^1 f(x, y) \, dx \, dy \approx J_1 = \tfrac{1}{4}\{f(b, b) + f(a, b) + f(a, a) + f(b, a)\}$$

$$a = \frac{1}{2}\left(1 + \frac{1}{\sqrt{3}}\right) \quad b = \frac{1}{2}\left(1 - \frac{1}{\sqrt{3}}\right).$$

Clearly these are only the simplest members of large families of such formulae (e.g. Abramowitz and Stegun, 1964).

4.5.2 Solving the Global Equations

For very small problems (say up to 50 nodes) the matrix A can be stored explicitly and the system $A\mathbf{U} = \mathbf{f}$ solved using a Crout factorization routine (or Cholesky if A is symmetric positive definite). A description of these basic solution routines can be found in, for example, Wait (1979b). This simple-minded approach soon becomes impractical as the size of the problem increases. A 10×10 mesh of finite element nodes yields a 100×100 matrix and as the mesh size is halved the number of nodes is increased by a factor of 4 and the storage required for the whole matrix by a factor of 16.

However, in the 100×100 matrix as nonzero components correspond to nodes in the same element, it follows that

$$a_{ij} = 0 \quad |i - j| \geqslant 12.$$

As a Crout factorization only effects components within this *band* (see Wait 1979b) it is not necessary to store zeros outside the band. If the problem is symmetric and positive definite, only half the band is necessary. It is invariably necessary to incorporate (partial) pivoting into the nonsymmetric factorization: this can increase the band width, but only by a factor of 2.

For problems in which the finite element grid is not so uniform, the band containing the nonzeros will not be of a constant width. Again, however, if it is not necessary to incorporate pivoting into the factorization, i.e. if the matrix is symmetric positive definite, then it is only necessary to store the *profile* of the matrix. In a finite element calculation, the profile will be symmetric even if the matrix values are not.

There are a number of ordering strategies that attempt to reduce (but not minimize) the storage required; such methods as *reverse Cuthill–McKee* and *nested dissection* are described in Wait (1979b).

If the problem is so large that the profile cannot be stored without the use of backing store then it may be advisable to consider an alternative solution strategy such as frontal solution or iteration.

Frontal solution

A frontal solution (or the frontal method: Irons, 1970), is simply Gauss elimination linked into the assembly strategy. If it is possible to store the matrix profile without having to resort to backing store then the frontal method is equivalent to a profile solver with a particular ordering of the nodes.

The shape and size of the profile is governed by the ordering of the nodes and in a frontal solution the ordering of the nodes is effectively determined by the ordering of the elements. Thus it follows that, if the frontal solver is to be used effectively, then the ordering of the elements must be done with care.

The frontal method is therefore a convenient form of profile solver when the problems are too large to be solved without the use of backing store, assuming again that there is no need to incorporate any pivoting into the algorithm.

Iteration

If the Galerkin equations are solved by a direct method (i.e. Gauss elimination, Cholesky factorization, etc.), whether by a frontal method or an explicit profile solver, it has to be accepted that room has to be found to store a large number of zeros. These are the zeros within the profile and which are replaced by nonzero values during the course of the elimination. It is possible to develop highly sophisticated algorithms that do not require the entire profile to be stored explicitly but we shall not discuss them here (see for example Wait, 1979b; George and Liu, 1981).

Typically, iterative methods are used to solve *finite difference* problems for these very reasons. But the theory of such methods and their rapid convergence relies on the properties of the matrix. For example successive over-relaxation (s.o.r.) assumes the matrix has 'property A' (see for example Forsythe and Wasow, 1960). Finite element matrices, particularly if the mesh is irregular or if the elements have side nodes, do not possess these desirable qualities. The convergence of s.o.r. for finite element problems is not guaranteed, but it is sometimes useful; see for example the numerical results of Chapter 5.

There are two other types of iteration that can prove useful in finite element calculations. These are alternating direction methods and conjugate gradients.

4.5.3 Alternating Direction Galerkin (ADG) methods for rectangular regions (Douglas and Dupont, 1971)

When finite element methods are applied to linear problems in one dimension, the resulting algebraic system has a simple *banded* from, whereas problems in more than one dimension give rise to *block banded* matrices in which each block is itself banded. For example in two dimensions a bilinear approximation often leads to a matrix of the form:

$$A = \begin{bmatrix} D & E & & & & \\ C & D & E & & & \\ & & \cdot & \cdot & \cdot & \\ & & & \cdot & \cdot & \cdot \\ & & & & \cdot & \cdot & \cdot \\ & & & & C & D & E \\ & & & & & C & D \end{bmatrix}$$

where C, D, and E are tridiagonal submatrices.

The object of ADG procedures in finite element methods is similar to that of ADI procedures in finite difference methods (Mitchell and Griffiths, 1980), namely to reduce the algebraic system derived from a multidimensional problem to a sequence of algebraic systems similar in form to those derived from one-dimensional problems.

In order to apply ADG methods it is necessary to assume that the basis functions are of the *tensor product form*, that is

$$\varphi_{ij}(x, y) = \varphi_i(x)\varphi_j(y) \quad (i = 1, \ldots, N_x; j = 1, \ldots, N_y).$$

The matrix $G = \{a(\varphi_{ij}, \varphi_{kl})\}$ can then be factorized such that

$$a(\varphi_{ij}, \varphi_{kl}) = a_x(\varphi_i, \varphi_k)b_y(\varphi_j, \varphi_l) + b_x(\varphi_i, \varphi_k)a_y(\varphi_j, \varphi_l)$$
$$(i, k = 1, \ldots, N_x; j, l = 1, \ldots, N_y).$$

If G is an $N \times N$ matrix and B is an $M \times M$ matrix the *matrix tensor product*, denoted by $G \otimes B$ is the $NM \times NM$ matrix:

$$\begin{bmatrix} g_{11}B & \cdots & g_{1N}B \\ \vdots & & \vdots \\ g_{N1}B & \cdots & g_{NN}B \end{bmatrix}.$$

The matrix can then be written as

$$A = G_x \otimes B_y + B_x \otimes G_y$$

if the nodes are ordered by columns.

If $R = [\alpha, \beta] \times [\gamma, \delta]$ and

$$a(\varphi_{ij}, \varphi_{kl}) = \int\int_R \left\{ \frac{\partial \varphi_{ij}}{\partial x} \frac{\partial \varphi_{kl}}{\partial x} + \frac{\partial \varphi_{ij}}{\partial y} \frac{\partial \varphi_{kl}}{\partial y} \right\} dxdy$$

it follows that

$$\{G_x\}_{ik} = g_x(\varphi_i, \varphi_k) = \int_\alpha^\beta \frac{d\varphi_i}{dx} \frac{d\varphi_k}{dx} dx$$

and

$$\{B_x\}_{ik} = b_x(\varphi_i, \varphi_k) = \int_\alpha^\beta \varphi_i \varphi_k \, dx.$$

Thus if $h = (\beta - \alpha)/(T + 1)$, with Dirichlet boundary conditions, it follows from (4.42) that in this example we have $T \times T$ matrices

$$B_x = \frac{h}{6} \begin{bmatrix} 4 & 1 & & & \\ 1 & 4 & 1 & & \\ & \cdot & \cdot & \cdot & \\ & & \cdot & \cdot & \cdot \\ & & & 1 & 4 & 1 \\ & & & & 1 & 4 \end{bmatrix} \qquad G_x = \frac{1}{h} \begin{bmatrix} 2 & -1 & & & \\ -1 & 2 & -1 & & \\ & \cdot & \cdot & \cdot & \\ & & \cdot & \cdot & \cdot \\ & & & -1 & 2 & -1 \\ & & & & -1 & 2 \end{bmatrix}$$

in the case of piecewise *linear* basis functions, and $(2T+1) \times (2T+1)$ matrices

$$B_x = \frac{h}{10} \begin{bmatrix} 8 & 1 & & & & & & & \\ 2 & 8 & 2 & -1 & & & & & \\ & 1 & 8 & 1 & & & & & \\ & -1 & 2 & 8 & 2 & -1 & & & \\ & & \cdot & \cdot & \cdot & \cdot & \cdot & \cdot & \\ & & & & -1 & 2 & 8 & 2 & -1 \\ & & & & & 1 & 8 & 1 & \\ & & & & & -1 & 2 & 8 & 2 \\ & & & & & & & 1 & 8 \end{bmatrix}$$

$$G_x = \frac{1}{h} \begin{bmatrix} 8 & -4 & & & & & & & \\ -8 & 14 & -8 & 1 & & & & & \\ & -4 & 8 & -4 & & & & & \\ & 1 & -8 & 14 & -8 & 1 & & & \\ & & \cdot & \cdot & \cdot & \cdot & \cdot & \cdot & \\ & & & & 1 & -8 & 14 & -8 & 1 \\ & & & & & -4 & 8 & -4 & \\ & & & & & 1 & -8 & 14 & -8 \\ & & & & & & & -4 & 8 \end{bmatrix}$$

for piecewise *quadratic* basis functions where *h is the element size.*
    The algebraic system can now be written as

$$\{G_x \otimes B_y + B_x \otimes G_y\}\alpha = \mathbf{b}$$

which is solved by means of the iteration

$$(\lambda_n B_x + G_x) \otimes (\lambda_n B_y + G_y)\alpha^{(n)}$$
$$= (\lambda_n B_x - G_x) \otimes (\lambda_n B_y - G_y)\alpha^{(n-1)} + 2\lambda_n \mathbf{b} = \psi^{(n-1)}$$

as a two-stage procedure,

$$(\lambda_n B_x + G_x) \otimes I_{N_y}\alpha^{(n^*)} = \psi^{(n-1)} \tag{4.102a}$$

and

$$I_{N_x} \otimes (\lambda_n B_y + G_y)\alpha^{(n)} = \alpha^{(n^*)} \tag{4.102b}$$

It is possible to split these equations such that if

$$\alpha_{p,C} = (\alpha_{p1}, \ldots, \alpha_{pN_y})^T \quad (p = 1, \ldots, N_x)$$

is a column of values on the grid and

$$\alpha_{p,R} = (\alpha_{1p}, \ldots, \alpha_{N_xp})^T \quad (p = 1, \ldots, N_y)$$

is a row of values on the grid, (4.102a) becomes

$$(\lambda_n B_x + G_x)\alpha_{p,R}^{(n^*)} = \psi_{p,R}^{(n-1)} \quad (p = 1, \ldots, N_y)$$

and (4.102b) becomes

$$(\lambda_n B_y + G_y)\alpha_{p,C}^{(n)} = \alpha_{p,C}^{(n^*)} \quad (p = 1, \ldots, N_x).$$

Douglas and Dupont have shown that, with a suitable choice for the sequence of iteration parameters $\{\lambda_n\}$, ADG is a rapidly convergent iterative procedure.

If the error in each iteration is

$$\mathbf{e}^{(n)} = \alpha^{(n)} - \alpha$$

then

$$\mathbf{e}^{(n)} = (B_x + G_x)^{-1}(\lambda_n B_x - G_x) \otimes (\lambda_n B_y + G_y)^{-1}(\lambda_n B_y - G_y)\mathbf{e}^{(n-1)}$$

as G_x, G_y are symmetric and B_x, B_y are symmetric positive definite. Writing

$$\varepsilon^{(n)} = B_x^{1/2} \otimes B_y^{1/2}\mathbf{e}^{(n)}$$

and

$$B^{-1/2}GB^{1/2} = C$$

this is equivalent to

$$\varepsilon^{(n)} = (\lambda_n + C_x)^{-1}(\lambda_n - C_x) \otimes (\lambda_n + C_y)^{-1}(\lambda_n - C_y)\varepsilon^{(n-1)}.$$

Thus if

$$C\mathbf{y} = \mu\mathbf{y}$$

then

$$(\lambda + C)^{-1}(\lambda - C)\mathbf{y} = \left(\frac{\lambda - \mu}{\lambda + \mu}\right)\mathbf{y}$$

and a sequence of parameters $\lambda_n, n = 1, \ldots, M$, leads to the bound

$$\| \varepsilon^{(M)} \| \leqslant K_M \| \varepsilon^{(0)} \|$$

where

$$K_M \leqslant \max_{\mu\epsilon\sigma(c)} \prod_{n=1}^{M} \left(\frac{\lambda_n - \mu}{\lambda_n + \mu}\right)$$

and $\sigma(C)$, the spectrum of C, contains all the eigenvalues of the matrix C.

Since

$$\mu_{\min} \geqslant \left(\frac{\pi}{\beta - \alpha}\right)^2 = \mu$$

and in general

$$\mu_{\max} \approx O(N^2) = \bar{\mu}$$

it follows that a realistic choice of parameters minimizes

$$\max_{x \in [\mu, \bar{\mu}]} \prod_{n=1}^{M} \left(\frac{\lambda_n - x}{\lambda_n + x} \right)$$

in a manner similar to the choice of Peaceman–Rachford parameters for ADI difference schemes (Peaceman and Rachford, 1955). It is possible to extend the formulation to include isoparametric quadrilaterals on a curved grid that is topologically equivalent to a rectangle if the Jacobians are replaced by a tensor product approximation (Hayes, 1981).

Exercise 4.15 What is the ADG formulation corresponding to the three-dimensional problem

$$\frac{\partial^2 u}{\partial x^2} + \frac{\partial^2 u}{\partial y^2} + \frac{\partial^2 u}{\partial z^2} = f(x, y, z)$$

on the parallelepiped $R = [a, b] \times [c, d] \times [e, g]$?

4.5.4 Conjugate gradients

An alternative iterative method that does not rely either on the selection of parameters or on the region being rectangular is the method of *preconditioned conjugate gradients*.

Using *exact* arithmetic this method can be used to solve any set of linear algebraic equations

$$A\mathbf{x} = \mathbf{b} \tag{4.103}$$

where A is $N \times N$, symmetric, and positive definite, in N iterations. An efficient implementation aims to get a good solution, using *finite* arithmetic, in considerably less than N iterations by clustering the eigenvalues in some way.

In general the eigenvalues of A will be evenly spread so it is necessary to *precondition* (4.103) in order to cluster the eigenvalues. The procedure is to find a matrix $\tilde{A} \equiv \tilde{L}\tilde{L}^T$ close to A and replace (4.103) by

$$(\tilde{L}^{-1} A \tilde{L}^T) \tilde{L}^T \mathbf{x} = \tilde{L}^{-1} \mathbf{b}$$

i.e.

$$\hat{A}\hat{\mathbf{x}} = \mathbf{b}.$$

If $\tilde{A}$ is close to A then $\hat{A}$ should be close to the identity and hence have eigenvalues clustered around unity.

The matrix $\tilde{A}$, or rather the factors $\tilde{L}\tilde{L}^T$, are found by *incomplete Cholesky factorization* of A. The reason for selecting an iterative method is to avoid the need to store the fill-in generated during a Cholesky factorization. The incomplete factorization involves ignoring most (or all) of the fill-in and so reducing the amount of additional storage required.

5

Time-dependent Problems

In Chapter 4 we derived approximate methods of solution for boundary value problems. In this chapter we extend such methods, where possible, to the solution of time-dependent initial value problems. Most of the methods discussed are based on a Galerkin discretization in space followed by a difference approximation of the derivatives in the resulting *semi-discrete* system of time-dependent ordinary differential equations. The Ritz method, introduced at the beginning of Chapter 4, is even less appropriate for time-dependent problems as new difficulties arise when considering the variational formulation of time-dependent problems (Mitchell, 1976). In dissipative systems, for example, the adjoint problem is introduced to provide a combined functional $I(u, u^*)$ and in evolutionary problems for which a true variational principle exists, such as Hamiltonian dynamics, the stationary value is *not* an extremum (see Chapter 2).

5.1 Hamilton's Principle

In Section 2.4 the equations of motion of a continuous dynamical system were derived as necessary conditions for a stationary point of a functional. We now show that if this formulation is used an approximate solution can be obtained as in Chapter 4, by determining a stationary value with respect to an approximating subspace of functions. As the equations of motion of such dynamical systems are of hyperbolic or parabolic type, it follows that the corresponding functional is not positive definite. Hence even for conservative systems the stationary value does *not* give an extremum and it is not possible to derive a best approximation. An example of such a functional corresponding to the equation of motion

$$\frac{\partial^2 u}{\partial t^2} - c^2 \frac{\partial^2 u}{\partial x^2} = 0 \tag{5.1}$$

for a vibrating string is

$$I(v) = \tfrac{1}{2}\rho \int_{t_0}^{t_1} \int_0^l \left[\left(\frac{\partial v}{\partial t} \right) - c^2 \left(\frac{\partial v}{\partial x} \right)^2 \right] \mathrm{d}x \mathrm{d}t. \tag{5.2}$$

Boundary conditions

It would appear that in order to obtain an approximate solution of the form

$$U(x, t) = \sum_{i=1}^N U_i \varphi_i(x, t)$$

to (5.1) we simply apply the Ritz method of Section 4.1 with the functional $I(v)$ given in (5.2); unfortunately this leads to inconsistencies in the boundary conditions.

The solution of a linear *hyperbolic* differential equation of the form

$$\frac{\partial^2 u(\mathbf{x}, t)}{\partial t^2} + A u(\mathbf{x}, t) = f(\mathbf{x}, t) \quad (t_0 < t \leqslant t_1) \tag{5.3}$$

for $\mathbf{x} \in R \subset \mathbb{R}^m$ with boundary ∂R, where A is a linear elliptic differential operator similar to those introduced in Chapter 4, is well defined in $R \times [t_0, t_1]$ if the conditions

$$u(\mathbf{x}, t) = 0 \qquad ((\mathbf{x}, t) \in \partial R \times (t_0, t_1)) \tag{5.3a}$$

$$u(\mathbf{x}, t_0) = u_0(\mathbf{x}) \qquad (\mathbf{x} \in R) \tag{5.3b}$$

and

$$\frac{\partial u(\mathbf{x}, t_0)}{\partial t} = v_0(\mathbf{x}) \qquad (\mathbf{x} \in R) \tag{5.3c}$$

hold. However the problem is not well-posed if (5.3c) is replaced by a condition at $t = t_1$, such as

$$u(\mathbf{x}, t_1) = u_1(\mathbf{x}) \quad (\mathbf{x} \in R). \tag{5.3d}$$

The problem of well-posed (or ill-posed) boundary conditions for hyperbolic problems is discussed in Section 7.6; it is related to the position of the boundary relative to the *characteristic curves* of the differential equation.

It is this problem of equation (5.3), subject to (5.3a), (5.3b), and (5.3d) that follows from a stationary value of the corresponding functional

$$I(v:t_0, t_1) = \int_{t_0}^{t_1} \{ (v_t, v_t) - a(v, v) + 2(f, v) \} \, \mathrm{d}t \tag{5.4}$$

where the inner product notation is as in Chapter 4.

If the function $u(\mathbf{x}, t)$ is the solution of (5.3), subject to the given initial and boundary conditions, then it follows that for any T_0 and T_1 such that $t_0 \leqslant T_0$

$< T_1 \leqslant t_1$ the function $u(\mathbf{x}, t)$ provides a stationary value of the integral $I(v: T_0, T_1)$ where all the admissible functions satisfy the conditions

$$v(\mathbf{x}, T_0) = u(\mathbf{x}, T_0) \quad (\mathbf{x} \in R) \tag{5.5a}$$

and

$$v(\mathbf{x}, T_1) = u(\mathbf{x}, T_1) \quad (\mathbf{x} \in R) \tag{5.5b}$$

together with the boundary condition

$$v(\mathbf{x}, t) = 0 \quad ((\mathbf{x}, t) \in \partial R \times (T_0, T_1)). \tag{5.6}$$

In particular, if we partition the interval $[t_0, t_1]$ as

$$t_0 = \tau_0 < \tau_1 \cdots < \tau_{K+1} = t_1.$$

then the solution of equation (5.3), together with the given initial and boundary conditions, provides a stationary value for each of the functionals

$$I_n(v) = \int_{\tau_{n-1}}^{\tau_{n+1}} \{(v_t, v_t) - a(v, v) + 2(f, v)\} \, \mathrm{d}t \quad (n = 1, \ldots, K) \tag{5.7}$$

As an example we describe one way of computing a finite element solution to equation (5.1). We subdivide the region $0 \leqslant x \leqslant l, \tau_{n-1} \leqslant t \leqslant \tau_{n+1}$, by means of the partition

$$0 = x_0 < x_1 < \cdots < x_{L+1} = l$$

where $x_{i+1} - x_i = h(i = 0, 1, 2, \ldots, L)$ and where we assume that

$$\tau_{n+1} - \tau_n = \tau_n - \tau_{n-1} = \Delta t.$$

Then we proceed as if to solve the boundary-value problem

$$\frac{\partial^2 u}{\partial t^2} - c^2 \frac{\partial^2 u}{\partial x^2} = 0 \quad (0 < x < l; \tau_{n-1} < t < \tau_{n+1}) \tag{5.8}$$

subject to

$$u(x, \tau_{n+1}) = u_{n+1}(x) \quad (0 \leqslant x \leqslant l) \tag{5.9a}$$

$$u(x, \tau_{n-1}) = u_{n-1}(x) \quad (0 \leqslant x \leqslant l) \tag{5.9b}$$

and

$$u(0, t) = u(l, t) = 0 \quad (\tau_{n-1} < t < \tau_{n+1}). \tag{5.9c}$$

In order to obtain an approximate solution of (5.8), subject to (5.9a), (5.9b), and (5.9c), we define in the region $(0 \leqslant x \leqslant l; \tau_{n-1} \leqslant t \leqslant \tau_{n+1})$ bilinear basis functions $\varphi_{ij}(x, t)(i = 1, \ldots, L; j = n - 1, n, n + 1)$ corresponding to the points $P_i^{(j)} = (x_i, \tau_j)$

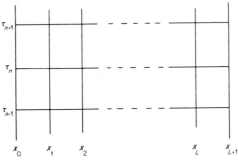

Figure 5.1

(cf. Section 4.2.2) as in Figure 5.1. The approximate solution is then of the form

$$U(x,t) = \sum_{j=n-1}^{n+1} \sum_{i=1}^{L} \varphi_{ij}(x,t) U_i^{(j)} \tag{5.10}$$

where $U_i^{(j)}$ is the approximate solution at the point P_i^j. For solving boundary-value problems defined by (5.8) – (5.9c), $U_i^{(n+1)}$ and $U_i^{(n-1)}$ ($i = 1,\ldots,L$) are determined from (5.9a) and (5.9b) respectively and U_i^n by

$$\frac{\partial}{\partial U_i^n} I_n\left(\sum_{j=n-1}^{n+1} \sum_{k=1}^{L} \varphi_{kj} U_k^{(j)}\right) = 0 \quad (i = 1,\ldots,L). \tag{5.11}$$

We are, however, attempting to solve an initial-value problem, not a boundary-value problem. Hence if U_i^{n-1} and U_i^n ($i = 1,\ldots,L$) are assumed known, (5.11) can be used to determine U_i^{n+1}.

Exercise 5.1 Verify that the step-by-step solution of the equation

$$\frac{\partial^2 u}{\partial t^2} - \frac{\partial^2 u}{\partial x^2} = 0 \quad (0 < x \leqslant l; t_0 < t \leqslant t_1)$$

as described above using bilinear basis functions leads to the system of difference equations

$$\{\delta_t^2 I_x - r^2 \delta_x^2 I_t\} U_i^{(n)} = 0 \tag{5.12}$$

where δ_t^2 is the second-order difference operator defined by

$$\delta_t^2 U_i^{(n)} = U_i^{(n+1)} - 2U_i^{(n)} + U_i^{(n-1)}$$

with δ_x^2 similarly, I_t is the Simpson's rule averaging operator

$$I_t U_i^{(n)} = \tfrac{1}{6}(U_i^{(n+1)} + 4U_i^{(n)} + U_i^{(n-1)})$$

with I_x similarly, and $r = \Delta t/h$ is known as the mesh ratio.

Initial conditions

In the example given above and in step-by-step methods in general derived similarly for (5.3)–(5.3c), it is necessary to approximate the initial conditions in some way to derive the approximate solution at $t = \tau_0$ and $t = \tau_1$, so that the functional $I_1(v)$ can be used to derive the solution at $t = \tau_2$. If there is no discontinuity between the initial conditions and the boundary conditions, then it is possible to define an approximate solution of the form

$$U(\mathbf{x}, t) = u_0(\mathbf{x}) + \sum_{i=1}^{N} U_i \varphi_i(\mathbf{x}, t). \tag{5.13}$$

This procedure is equivalent to transforming the problem (cf. Section 4.2) to one for which the initial condition (5.3b) is replaced by

$$U(\mathbf{x}, t_0) = 0 \quad (\mathbf{x} \in R).$$

When there is a discontinuity, that is,

$$u_0(\mathbf{x}) \neq 0 \quad (\mathbf{x} \in \partial R)$$

it is not possible to use an approximation of the form (5.13) and it is necessary to approximate $u_0(\mathbf{x})$ in terms of

$$U(\mathbf{x}, t_0) = \sum_{i=1}^{N} U_i \varphi_i(\mathbf{x}, t_0)$$

such that either $U(\mathbf{x}, t_0)$ interpolates $u_0(\mathbf{x})$ or the error $u_0(\mathbf{x}) - U(\mathbf{x}, t_0)$ is minimized in some norm. Similarly, it is necessary to approximate the second initial condition so that

$$\frac{\partial U(\mathbf{x}, t_0)}{\partial t} = \sum_{i=1}^{N} U_i \frac{\partial \varphi_i(\mathbf{x}, t_0)}{\partial t}$$

is a good approximation to $v_0(\mathbf{x})$.

Thus in the earlier example, it is possible to introduce the initial conditions

$$U_i^{(0)} = u_0(x_i) \quad (i = 1, \dots, L)$$

and

$$\frac{U_i^{(1)} - U_i^{(0)}}{\Delta t} = v_0(x_i) \quad (i = 1, \dots, L).$$

This use of Hamilton's principle leads to a valid step-by-step method of solution for conservative systems, although the mathematical formulation of such a method is in parts confusing. Similar procedures describing step-by-step methods of approximation for hyperbolic equations are given by Noble (1973) and Mitchell (1976).

5.1.1 Dissipative systems

As we mentioned at the beginning of this chapter, various authors have attempted to provide a variational formulation for dissipative problems. Some have attempted to provide a general variational principle that holds for a large class of such problems (Finlayson and Scriven, 1967, and references therein) as Hamilton's principle does for conservative systems. The main drawbacks of such formulations are that:

(1) The variational principles are derived from the constitutive equations rather than the other way round, hence the variational formulation involves additional effort, but provides no additional information.
(2) The functional involved in the variational principle has no physical significance, and the stationary values are never true extrema that can be used to derive error bounds.
(3) As indicated in the previous section there are basic discrepancies between variational problems and initial-value problems that are invariably ignored in most so-called variational formulations of evaluationary problems.

We have stressed the merits and disadvantages of variational formulations at this point in the book, in view of the wide variety of such formulations of *dissipative* or *irreversible* processes in the literature, and the often contradictory claims made about them. We now introduce the finite element solution of evolutionary dissipative systems using the adjoint formulation by providing two exercises which may be attempted by the interested reader.

Other numerical solutions based on variational treatments of the diffusion equation can be found in Noble (1973), Cecchi and Cella (1973), and Mitchell (1976).

N.B. A final discouraging note regarding the use of finite element methods based on variational principles for evolutionary problems is that the elements are in space and time and so, for example, three-dimensional elements are required for a two-space-dimensional problem.

Exercise 5.2 Verify, after consulting Chapter 2 and suitable tests, that the functional corresponding to the simple diffusion equation

$$\frac{\partial u}{\partial t} = \frac{\partial^2 u}{\partial x^2} \quad (0 < x < l; t_0 < t \leqslant t_1)$$

subject to

$$u(0, t) = u(l, t) = 0 \quad (t_0 < t \leqslant t_1)$$

together with

$$u(x, t_0) = u_0(x) \quad (0 \leqslant x \leqslant l)$$

is given by

$$I = \int_{t_0}^{t_1} \int_0^1 \left\{ \frac{\partial v}{\partial t} v^* + \frac{\partial v}{\partial x} \frac{\partial v^*}{\partial x} \right\} dx dt.$$

Then show that the equation

$$-\frac{\partial u^*}{\partial t} = \frac{\partial^2 u^*}{\partial x^2}$$

is the necessary condition corresponding to the stationary point $\delta_v I(v, v^*) = 0$, if either (i) u is given for $t = t_0$ and $t = t_1$ or (ii) u is given for $t = t_0$ and $u^* = 0$ for $t = t_1$.

Note This shows that when calculating a stationary point of I_n, we should assume that $u(x, \tau_{n-1})$ can be any given value, whereas $u^*(x, \tau_n) = 0$. Thus the function $U^*(x, t)$ is not the approximate solution to a single adjoint system but is rather a sequence of approximations to separate adjoint systems corresponding to each interval $[\tau_{n-1}, \tau_n]$, each with the final condition

$$U^*(x, \tau_n) = 0 \quad (n = 1, 2, \ldots, K)$$

for $0 \leqslant x \leqslant l$.

Exercise 5.3 Using Exercise 5.2, show that the solution of the diffusion equation

$$\frac{\partial u}{\partial t} - \frac{\partial^2 u}{\partial x^2} = 0 \quad (0 < x < l; t_0 < t \leqslant t_1)$$

with bilinear basis functions, leads to the system of difference equations

$$\Delta_t I_x U_i^{(n)} - \tfrac{1}{3} r \delta_x^2 \{ U_i^{(n)} + 2 U_i^{(n+1)} \} = 0 \tag{5.14}$$

where Δ_t is the forward difference operator in t; I_x, δ_x^2 are difference operators in x as defined earlier, and where $r = \Delta t/(\Delta x)^2$. The region is partitioned such that $\tau_{n+1} - \tau_n = \Delta t (n = 0, 1, \ldots, K)$ and $x_{i+1} - x_i = \Delta x$ $(i = 0, 1, \ldots, N)$. This particular time-stepping formula (5.14) appears several times in this chapter, and is generated by a number of different formulations of finite element approximations (cf. (5.35a) and (5.45)).

5.2 Semi-discrete Galerkin Methods

5.2.1 Introduction

Semi-discrete methods are techniques for by-passing variational formulations for evolutionary problems. Discretization is carried out in the space variables only, usually by a Galerkin method, leading to a system of ordinary differential equations in time.

As an example, consider the differential equation

$$\frac{\partial u(\mathbf{x}, t)}{\partial t} + Au(\mathbf{x}, t) = f(\mathbf{x}, t) \quad ((\mathbf{x}, t) \in R \times (t_0, t_1]) \tag{5.15}$$

where A is a second-order differential operator such as, in two dimensions,

$$A = -\frac{\partial^2}{\partial x^2} - \frac{\partial^2}{\partial y^2}.$$

A solution of (5.15) is required, subject to the initial condition

$$u(\mathbf{x}, t_0) = u_0(\mathbf{x}) \quad (\mathbf{x} \in R) \tag{5.15a}$$

and the boundary condition

$$u(\mathbf{x}, t) = 0 \quad ((\mathbf{x}, t) \in \partial R \times (t_0, t_1]). \tag{5.15b}$$

The corresponding weak form of the problem is, in the notation of Section 4.4, that for $t \in (t_0, t_1]$

$$\left(\frac{\partial u}{\partial t}, v\right) + a(u, v) = (f, v) \quad \text{(for all } v(\mathbf{x}) \in \mathcal{H}) \tag{5.16}$$

subject to the initial condition

$$(u, v)_{t = t_0} = (u_0, v) \quad \text{(for all } v(\mathbf{x}) \in \mathcal{H}). \tag{5.16a}$$

In this model problem it is clear that—using Sobolev space notation—$\mathcal{H} = \mathring{\mathcal{H}}^{(1)}(R)$ and also that $u \in \mathcal{H} \times C^1[t_0, t_1]$. If a more general boundary condition is specified, then it may be necessary to modify the weak form by the addition of boundary integrals. Further details of this method of modifying functionals is given in Chapters 2, 4, and 6.

The semi-discrete approximation U is then defined in terms of the weak form of the equation; that is, for $t \in (t_0, t_1]$

$$\left(\frac{\partial U}{\partial t}, V\right) + a(U, V) = (f, V) \quad \text{(for all } V(\mathbf{x}) \in K_N) \tag{5.17}$$

subject to the initial condition

$$(U, V)_{t = t_0} = (u_0, V) \quad \text{(for all } V(\mathbf{x}) \in K_N). \tag{5.17a}$$

For the model problem it follows that $K_N \subset \mathring{\mathcal{H}}^{(1)}(R)$. If the functions $\varphi_i (i = 1, \ldots, N)$ form a basis for the subspace K_N, the equivalent formulation of the semi-discrete approximation is that for $t \in (t_0, t_1]$

$$\left(\frac{\partial U}{\partial t}, \varphi_i\right) + a(U, \varphi_i) = (f, \varphi_i) \quad (i = 1, \ldots, N) \tag{5.18}$$

$$(U, \varphi_i)_{t = t_0} = (u_0, \varphi_i) \quad (i = 1, \ldots, N) \tag{5.18a}$$

where $U \in K_N \times C^1[t_0, t_1]$, the Galerkin approximation is of the form

$$U(\mathbf{x}, t) = \sum_{i=1}^{N} U_i(t)\varphi_i(\mathbf{x}). \tag{5.19}$$

If the boundary conditions are inhomogeneous Dirichlet conditions, then it is possible to define a Galerkin approximation of the form

$$U(\mathbf{x}, t) = W(\mathbf{x}, t) + \sum_{i=1}^{N} U_i(t)\varphi_i(\mathbf{x})$$

where $\varphi_i \in K_N$, and $W(\mathbf{x}, t)$ satisfies the boundary conditions. It follows that, as for elliptic problems, it is only necessary for V to be in the energy space and no such requirement is placed on the approximation U.

The Galerkin approximation is defined by a system of ordinary differential equations in terms of the functions $U_i(t)$ $(i = 1, \ldots, N)$. It follows from (5.18) that, for the model problem, these equations can be written as

$$\sum_{j=1}^{N} \left\{ \frac{dU_i}{dt}(\varphi_j, \varphi_i) + U_j a(\varphi_j, \varphi_i) \right\} = (f, \varphi_i) \quad (i = 1, \ldots, N) \tag{5.20}$$

and the initial condition (5.18a) becomes

$$U_j(t_0) = c_j \quad (j = 1, \ldots, N) \tag{5.20a}$$

where

$$\sum_{j=1}^{N} (\varphi_j, \varphi_i) c_j = (u_0, \varphi_i) \quad (i = 1, \ldots, N). \tag{5.20b}$$

The coefficients $c_j (j = 1, \ldots, N)$ given by (5.20b) satisfy

$$\left\| u_0(\mathbf{x}) - \sum_{j=1}^{N} c_j \varphi_j(\mathbf{x}) \right\|_{0,R}^2 = \text{minimum}$$

and so in certain problems it might be appropriate to replace (5.20b) by a different approximation of the original data, or possibly alter the form of the approximation to satisfy the initial condition exactly.

In terms of the *mass matrix*

$$B = \{(\varphi_j, \varphi_i)\}$$

and the *stiffness matrix*

$$G = \{a(\varphi_j, \varphi_i)\}$$

the system (5.20) can be written as

$$B\dot{U} + GU = \mathbf{b} \tag{5.21}$$

where

$$\mathbf{U} = [U_1(t), \ldots, U_N(t)]^{\mathrm{T}}$$
$$\mathbf{b} = [f_1, \ldots, f_N]^{\mathrm{T}}$$

with

$$f_i = (f, \varphi_i) \quad (i = 1, \ldots, N)$$

and the dot represents differentiation with respect to t.

The initial condition (5.20b) can be written similarly as

$$B U(t_0) = \mathbf{u}^* \tag{5.21a}$$

where

$$u_i^* = (u_0, \varphi_i) \quad (i = 1, \ldots, N).$$

The procedure for incorporating boundary conditions other than homogeneous Dirichlet conditions is similar to that adopted in Chapter 4 for elliptic problems. Also the use of basis functions (e.g. cubic B-splines), where the nodal parameters are not function values, further complicates the system of ordinary differential equations (5.21).

It is perhaps worth pointing out that advocates of finite difference methods usually prefer to discretize in space, in a time-dependent problem, by the method of lines (Zafarullah, 1970) rather than to use a Galerkin-type procedure. The reasoning behind this is that if the mass matrix B in (5.21) is replaced by the unit matrix I (*mass-lumping*), then the system of ordinary differential equations is easier to solve. Our experience, particularly in nonlinear problems, has been that the spreading of the mass by Galerkin-type methods often has a beneficial effect on the solution for a small increase in computational effort.

5.2.2 Simple example

As an illustration of this method, consider the simple diffusion equation

$$\frac{\partial u}{\partial t} = \frac{\partial^2 u}{\partial x^2} \quad (0 < x < l; t > 0)$$

subject to

$$u(0, t) = u(l, t) = 0 \quad (t \geqslant 0)$$

and

$$u(x, 0) = u_0(x) \quad (0 < x < l).$$

The approximate solution is

$$U(x, t) = \sum_{i=1}^{N} u_i(t)\varphi_i(x)$$

where the basis functions satisfy the boundary condition and hence

$$\varphi_i(0) = \varphi_i(l) = 0 \quad (i = 1, \ldots, N).$$

The system of equations (5.20) then becomes

$$\sum_{j=1}^{N} \left\{ \frac{dU_j}{dt} b_{ij} + U_j g_{ij} \right\} = 0 \quad (i = 1, \ldots, N) \tag{5.22}$$

where

$$g_{ij} = \int_0^l \frac{d\varphi_i}{dx} \frac{d\varphi_j}{dx} dx \quad (i, j = 1, 2, \ldots, N) \tag{5.22a}$$

and

$$b_{ij} = \int_0^l \varphi_i \varphi_j \, dx \quad (i, j = 1, 2, \ldots, N). \tag{5.22b}$$

It follows from (5.22) that, in this example, equation (5.21) has

$$B = \frac{h}{6} \begin{bmatrix} 4 & 1 & & & \\ 1 & 4 & 1 & & \\ & \cdot & \cdot & \cdot & \\ & & \cdot & \cdot & 1 \\ & & & 1 & 4 \end{bmatrix} \qquad G = \frac{1}{h} \begin{bmatrix} 2 & -1 & & & \\ -1 & 2 & -1 & & \\ & \cdot & \cdot & \cdot & \\ & & \cdot & \cdot & -1 \\ & & & -1 & 2 \end{bmatrix}$$

in the case of piecewise *linear* basis functions, and

$$B = \frac{h}{10} \begin{bmatrix} 8 & 1 & & & & & & \\ 2 & 8 & 2 & -1 & & & & \\ & 1 & 8 & 1 & & & & \\ & -1 & 2 & 8 & 2 & -1 & & \\ & & & \cdot & \cdot & \cdot & \cdot & \cdot \\ & & & & -1 & 2 & 8 & 2 & -1 \\ & & & & & 1 & 8 & 1 \\ & & & & & -1 & 2 & 8 & 2 \\ & & & & & & & 1 & 8 \end{bmatrix}$$

$$G = \frac{1}{h} \begin{bmatrix} 8 & -4 & & & & & & \\ -8 & 14 & -8 & 1 & & & & \\ & -4 & 8 & -4 & & & & \\ & 1 & -8 & 14 & -8 & 1 & & \\ & & \cdot & \cdot & \cdot & \cdot & \cdot & \cdot \\ & & & & 1 & -8 & 14 & -8 & 1 \\ & & & & & -4 & 8 & -4 \\ & & & & & 1 & -8 & 14 & -8 \\ & & & & & & & -4 & 8 \end{bmatrix}$$

for piecewise *quadratic* basis functions where *h is the element size*. In the latter case, the unknowns are alternately at the half-integer and integer nodes (see Chapter 1).

5.2.3 Hyperbolic equations

Although semi-discrete methods have been studied theoretically, primarily in relation to parabolic equations, they can equally well be applied to hyperbolic equations, for example equations of the form

$$\frac{\partial^2 u}{\partial t^2} + \lambda \frac{\partial u}{\partial t} + Au = 0 \quad (\lambda > 0) \quad A = -\frac{\partial^2}{\partial x^2} - \frac{\partial^2}{\partial y^2}$$

that represent damped mechanical vibrations. In such problems a semi-discrete approximation of the form

$$U(\mathbf{x}, t) = \sum_{i=1}^{N} U_i(t) \varphi_i(\mathbf{x})$$

leads to a system of second-order ordinary differential equations

$$B(\ddot{U} + \lambda \dot{U}) + GU = 0. \tag{5.23}$$

There are two sets of initial conditions

$$U_i(t_0) = c_i \quad (i = 1, \ldots, N)$$

and

$$\frac{dU_i(t_0)}{dt} = d_i \quad (i = 1, \ldots, N)$$

the constants c_i and d_i being determined from the given initial conditions

$$u(\mathbf{x}, t_0) = u_0(\mathbf{x})$$

and

$$\frac{\partial u(\mathbf{x}, t_0)}{\partial t} = v_0(\mathbf{x})$$

such that

$$\left\| u_0(\mathbf{x}) - \sum_{i=1}^{N} c_j \varphi_i(\mathbf{x}) \right\|_{0,R}^2 = \text{minimum}$$

and

$$\left\| v_0(\mathbf{x}) - \sum_{i=1}^{N} d_i \varphi_i(\mathbf{x}) \right\|_{0,R}^2 = \text{minimum}$$

respectively. Thus the initial conditions can be written as

$$BU(t_0) = \mathbf{u}^*$$

$$B\dot{U}(t_0) = \mathbf{v}^*$$

where

$$u_i^* = (u_0, \varphi_i) \quad \text{and} \quad v_i^* = (v_0, \varphi_i).$$

Some progress has also been made in the application of semi-discrete Galerkin methods to the solution of *first*-order hyperbolic equations and systems; material involving first-order hyperbolic operators appears in Chapter 7.

5.2.4 Nonlinear problems

In Section 4.4 we gave an example of a nonlinear equation, $A(u) = f$, that can be solved by Galerkin's method. At the same time it is pointed out that the advantages of Galerkin's method are not fully realized if it is not possible to apply Green's theorem to simplify the inner products. The situation is very similar when we apply the semi-discrete Galerkin method to solve the nonlinear parabolic equation

$$\frac{\partial u}{\partial t} + A(u) = 0.$$

We consider, as an example, the same nonlinear term as in the elliptic case, namely

$$A(u) = -\frac{\partial}{\partial x}\left(p(u)\frac{\partial u}{\partial x}\right) - \frac{\partial}{\partial y}\left(q(u)\frac{\partial u}{\partial y}\right).$$

In order to apply the semi-discrete Galerkin method to this equation it is first necessary to rewrite the equation in the weak form (cf. (4.80))

$$\left(\frac{\partial u}{\partial t}, v\right) + a(u, v; u) = 0 \quad \text{(for all } v \in \mathscr{H})$$

where

$$a(u, v; w) = \int\int_R \left\{p(w)\frac{\partial u}{\partial x}\frac{\partial v}{\partial x} + q(w)\frac{\partial u}{\partial y}\frac{\partial v}{\partial y}\right\}\mathrm{d}x\mathrm{d}y$$

which leads to a semi-discrete approximation

$$U(\mathbf{x}, t) = \sum_j U_j(t)\varphi_j(\mathbf{x})$$

that satisfies

$$\left(\frac{\partial U}{\partial t}, \varphi_i\right) + a(U, \varphi_i; U) = 0 \quad (i = 1, \dots, N). \tag{5.24}$$

In terms of the parameters $\mathbf{U} = [U_1(t), \dots, U_N(t)]^{\mathrm{T}}$ this can be written as the system of nonlinear ordinary differential equations

$$\sum_{j=1}^N \frac{\mathrm{d}U_j}{\mathrm{d}t} b_{ij} + g_i(\mathbf{U}) = 0 \quad (i = 1, 2, \dots, N) \tag{5.24a}$$

This in turn can be written as

$$B\dot{\mathbf{U}} + \mathbf{g}(\mathbf{U}) = 0 \tag{5.24b}$$

where B is the mass matrix. The problem of solving such nonlinear systems is discussed later in this chapter. Other important nonlinear equations which can be solved by semi-discrete Galerkin methods are dealt with in Chapter 7.

5.3 Discretization in Time

5.3.1 Finite difference methods

In the previous section, the time-dependent problem was discretized in space using the Galerkin method resulting in a system of ordinary differential equations in time, the form of the latter depending on the original problem. A comprehensive study of numerical methods based on finite differences for general systems of ordinary differential equations has been provided elsewhere (Gear, 1971; Lambert, 1973). Nevertheless, the structure of a system obtained from a time-dependent partial differential equation should not be ignored. In particular for small element sizes in space, the system is likely to be *stiff* (Lambert, 1973, p. 231) and so only special methods of solution will be satisfactory in such a case (Laurie, 1976; Hopkins and Wait, 1978).

Parabolic partial differential equations and the resulting first-order (in time) systems have probably received most attention, the most popular method of solution being the so-called *Crank–Nicolson–Galerkin* (CNG) method. In this form of approximation, the system of differential equations (5.21) is replaced by the system of difference equations

$$B\left\{\frac{\mathbf{U}^{(n+1)} - \mathbf{U}^{(n)}}{\Delta t}\right\} + G\left\{\frac{\mathbf{U}^{(n+1)} + \mathbf{U}^{(n)}}{2}\right\} = \mathbf{b}(\tau_{n+1/2}) \quad (n = 0, 1, \ldots) \quad (5.25)$$

where $\mathbf{U}^{(n)}$ is an approximation to $\mathbf{u}(t_0 + n\Delta t)$ and $\tau_{n+1/2} = t_0 + (n + \frac{1}{2})\Delta t$ ($n = 0, 1, \ldots$). This form of approximate solution of a system of ordinary differential equations should be described more accurately as the trapezium method. From (5.25) it is clear that at each step of the calculation it is necessary to solve a system of linear algebraic equations to find the values of $\mathbf{U}^{(n+1)}$.

An alternative formulation of (5.25) is to consider the fully-discrete Galerkin equations

$$\left(\frac{U^{(n+1)} - U^{(n)}}{\Delta t}, \varphi_i\right) + a\left(\frac{U^{(n+1)} + U^{(n)}}{2}, \varphi_i\right) = (f^{(n+1/2)}, \varphi_i) \quad \begin{Bmatrix} i = 1, \ldots, N \\ n = 0, 1, \ldots \end{Bmatrix}$$

$$(5.26a)$$

as defining a sequence of approximations

$$U^{(n)}(\mathbf{x}) = \sum_{j=1}^{N} U_j^{(n)} \varphi_j(\mathbf{x}) \quad n = 0, 1, \ldots$$

such that $U^{(n)}(\mathbf{x})$ is an approximation to $U(\mathbf{x}, n\Delta t)$ where $U(\mathbf{x}, t)$ is the solution of the semi-discrete equations (5.18) and in (5.26a),

$$f^{(n+1/2)} = \begin{cases} f(\mathbf{x}, (n + \frac{1}{2})\Delta t) \\ \text{or} \\ \frac{1}{2}(f(\mathbf{x}, n\Delta t) + f(\mathbf{x}, (n + 1)\Delta t)). \end{cases}$$

As (5.26a) is linear in $U^{(n+1)}$ and $U^{(n)}$, it can be written as

$$\sum_{j=1}^{N} \left\{ \frac{U_j^{(n+1)} - U_j^{(n)}}{\Delta t} (\varphi_j, \varphi_i) + \frac{U_j^{(n+1)} + U_j^{(n)}}{2} a(\varphi_j, \varphi_i) \right\}$$

$$= (f^{(n+1/2)}, \varphi_i) \quad \begin{cases} i = 1, \ldots, N \\ n = 0, 1, \ldots \end{cases}. \tag{5.26b}$$

Given the definitions of the mass matrix B and the stiffness matrix G, this is clearly equivalent to (5.25).

Unfortunately the situation is not so simple for nonlinear equations of the form

$$B\dot{U} + g(U) = b \tag{5.27}$$

which arise from the semi-discrete Galerkin solution of nonlinear equations of the form

$$\frac{\partial u}{\partial t} + A(u) = f. \tag{5.28}$$

A CNG approximation in the case of (5.27) would lead to a system of nonlinear equations to solve for $U^{(n+1)}$, thus a *predictor–corrector* method has to be used. Douglas and Dupont (1970) suggest various schemes and discuss their relative merits, but it is only possible here to mention, briefly, a typical example of the methods they suggest for solving the equation (5.28) with

$$A(u) = -\frac{\partial}{\partial x}\left(p(u) \frac{\partial u}{\partial x} \right) - \frac{\partial}{\partial y}\left(q(u) \frac{\partial u}{\partial y} \right). \tag{5.29}$$

In this case we define the nonlinear stiffness matrix as

$$G(U) = \left\{ \iint_R \left\{ p(U) \frac{\partial \varphi_j}{\partial x} \frac{\partial \varphi_i}{\partial x} + q(U) \frac{\partial \varphi_j}{\partial y} \frac{\partial \varphi_i}{\partial y} \right\} dxdy \right\} \quad (i, j = 1, \ldots, N)$$

$$= \{a(\varphi_j, \varphi_i; U)\}. \tag{5.30}$$

Thus $g(U)$ can be written as

$$g(U) = G(U)U$$

and (6.34) becomes

$$B\dot{U} + G(U)U = b$$

and (5.25) is replaced by the two equations,

$$B\left\{ \frac{V^{(n+1)} - U^{(n)}}{\Delta t} \right\} + G(U^{(n)})\left\{ \frac{V^{(n+1)} + U^{(n)}}{2} \right\} = b(\tau_{n+1/2}) \tag{5.31a}$$

and

$$B\left\{ \frac{U^{(n+1)} - U^{(n)}}{\Delta t} \right\} + G\left(\frac{V^{(n+1)} + U^{(n)}}{2} \right)\left\{ \frac{U^{(n+1)} + U^{(n)}}{2} \right\} = b(\tau_{n+1/2}). \tag{5.31b}$$

The predictor (5.31a) gives a first approximation $\mathbf{V}^{(n+1)}$, then the corrector (5.31b) may be used iteratively to improve the approximation.

Once again the fully-discrete equations can be derived direct from the Galerkin form. In this case (5.24) can be discretized as

$$\left(\frac{U^{(n+1)} - U^{(n)}}{\Delta t}, \varphi_i\right) + a\left(\frac{U^{(n+1)} + U^{(n)}}{2}, \varphi_i; W^{(n)}\right) = (f^{(n+1/2)}, \varphi_i)$$

in terms of $U^{(n)}(\mathbf{x})$, $U^{(n+1)}(\mathbf{x})$, and $W^{(n)}(\mathbf{x})$. If $W^{(n)} = U^{(n)}$, the equation is linear in $U^{(n+1)}$ whereas if $W^{(n)} = \frac{1}{2}(U^{(n)} + U^{(n+1)})$ it is nonlinear. From the definitions it is clear that this fully-discrete Galerkin equation can be written as

$$B\left(\frac{\mathbf{U}^{(n+1)} - \mathbf{U}^{(n)}}{\Delta t}\right) + G(\mathbf{W}^{(n)})\frac{\mathbf{U}^{(n+1)} + \mathbf{U}^{(n)}}{2} = \mathbf{b}(\tau_{n+1/2}).$$

The predictor–corrector pair follows directly by first substituting $\mathbf{W}^{(n)} = \mathbf{U}^{(n)}$ in the predictor and then with $\mathbf{W}^{(n)} = \frac{1}{2}(\mathbf{U}^{(n+1)} + \mathbf{U}^{(n)})$ in the corrector. In terms of the Galerkin form, (5.31a) and (5.31b) could be written as

$$\left(\frac{V^{(n+1)} - U^{(n)}}{\Delta t}, \varphi_i\right) + a\left(\frac{V^{(n+1)} + U^{(n)}}{2}, \varphi_i; U^{(n)}\right) = (f^{(n+1/2)}, \varphi_i) \quad (5.32a)$$

and

$$\left(\frac{U^{(n+1)} - U^{(n)}}{\Delta t}, \varphi_i\right) + a\left(\frac{U^{(n+1)} + U^{(n)}}{2}, \varphi_i; \frac{V^{(n+1)} + U^{(n)}}{2}\right) = (f^{(n+1/2)}, \varphi_i).$$

$$(5.32b)$$

As in Section 4.4, it is usual for the nonlinear term in $a(u, v; w)$ to be linearized by product integration. A generalisation of the CNG method is obtained by using

$$\theta U^{(n+1)} + (1 - \theta)U^{(n)} \quad 0 \leqslant \theta \leqslant 1$$

instead of $\frac{1}{2}(\mathbf{U}^{(n+1)} + \mathbf{U}^{(n)})$ in (5.25). This leads to (using ordinary differential equation terminology) Euler's method when $\theta = 0$, the trapezoidal rule when $\theta = \frac{1}{2}$, and the backward Euler method when $\theta = 1$.

Exercise 5.4 Verify that the semi-discrete piecewise linear approximation to the wave equation

$$\frac{\partial^2 u}{\partial t^2} - c^2 \frac{\partial^2 u}{\partial x^2} = 0 \quad \begin{cases} 0 < x < l \\ t_0 < t \leqslant t_1 \end{cases}$$

can be discretized to give (5.12) if $\ddot{U}_i(\tau_n)$ is replaced by

$$\left(\frac{1}{\Delta t}\right)^2 \delta_t^2 U_i^{(n)} \equiv \left(\frac{1}{\Delta t}\right)^2 (U_i^{(n+1)} - 2U_i^{(n)} + U_i^{(n-1)})$$

and $U_i(\tau_n)$ is averaged as

$$I_t U_i^{(n)} \equiv \frac{1}{6}(U_i^{(n+1)} + 4U_i^{(n)} + U_i^{(n-1)}).$$

5.3.2 Numerical example

As an illustration consider the following numerical example:

$$\frac{\partial u}{\partial t} = \frac{\partial^2 u}{\partial x^2} + \frac{\partial^2 u}{\partial y^2} + 2\alpha \frac{\partial u}{\partial y} \qquad \begin{cases} |y| \leqslant \frac{1}{2}\pi \\ 0 \leqslant x \leqslant \frac{1}{2}\pi \\ t \geqslant 0 \end{cases} \qquad (5.33)$$

with the boundary conditions

$$u = 0 \qquad x = \tfrac{1}{2}\pi \quad \text{or} \quad y = -\tfrac{1}{2}\pi$$

$$\frac{\partial u}{\partial x} = 0 \qquad x = 0$$

(a symmetry condition: cf. numerical example 2 in Chapter 4) and

$$u = (\tfrac{1}{2}\pi)^2 - x^2 \qquad y = \tfrac{1}{2}\pi$$

together with the initial condition

$$u(x, y, 0) = 0 \qquad \begin{cases} |y| \leqslant \frac{1}{2}\pi \\ 0 \leqslant x \leqslant \pi. \end{cases} \qquad (5.34)$$

The problem is to be solved using piecewise linear triangles, interpolating the boundary condition on $y = \frac{1}{2}\pi$.

The solution tends to a steady state that is (4.50), the solution to the numerical example 2 of Chapter 4. The semi-discrete Galerkin equations

$$\sum_{j=1}^{N} \dot{U}_j \iint_R \varphi_i \varphi_j \, dx \, dy$$

$$+ \sum_{j=1}^{N} U_j \iint_R \left\{ \frac{\partial \varphi_i}{\partial x} \frac{\partial \varphi_j}{\partial x} + \frac{\partial \varphi_i}{\partial y} \frac{\partial \varphi_j}{\partial y} - 2\alpha \varphi_i \frac{\partial \varphi_j}{\partial y} \right\} dx \, dy = 0 \quad i = 1, \dots, N$$

are solved using the theta method.

Consider first the solution with $\alpha = 1$ and a grid of uniform triangles with $h = \frac{1}{8}\pi$. The results can then be compared directly with Table 4.4 and Figure 4.4.

Given the initial condition (5.34), the initial state of the numerical solution is shown in Figure 5.2.

The problem was then solved with $\theta = \frac{1}{2}$ (i.e. the CNG method) and a time step of 0.1. The numerical solutions after 2, 4, and 8 time steps are shown in Figure 5.3. The equations were solved using s.o.r. with an acceleration parameter $w = 1.5$ and the number of iterations for the individual time step is given. In each case the iteration was terminated when the residual $\mathbf{r}$ satisfied

$$\|\mathbf{r}\|_2 \leqslant 0.01$$

and the solution at the preceding time step was used as a first approximation.

TIME STEP 0   TIME 0

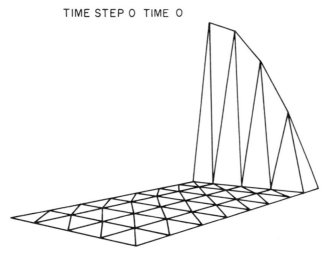

Figure 5.2 The initial state with a uniform grid of triangles

It can be clearly seen from Figure 5.3 that the solution exhibits the oscillations that characterize the CNG method. If we use a time step of 0.05 they are no longer present for this grid, whereas if we use 0.2 they are greatly exaggerated.

In addition it can be seen from Table 5.1 that after only 30 time steps the solution is getting very close to the steady state given in Table 4.4. If on the other hand the backward Euler method ($\theta = 1$) is used, there are no such ripples and the solution is very similar (see Figure 5.4 and Table 5.1).

As an alternative illustration consider the solution for $\alpha = 4$ together with the nonuniform grid illustrated in Figure 4.4. The CNG solution is shown in Figure 5.5: for this example the s.o.r. parameter was $w = 1.0$ (i.e. Gauss–Seidel), $w = 1.5$ did not converge, and once again there are oscillations.

For either problem, a value of $\theta < \frac{1}{2}$ leads to an unstable scheme (see Figure 5.6; $\theta = 0.4$, $\alpha = 4$). The difference between the oscillations with the CNG method and the unstable method are that from Figures 5.3 and 5.5 it can be seen that the CNG

Table 5.1 Solution after 30 time steps

Point	$\theta = \frac{1}{2}$	$\theta = 1$
1	1.804	1.814
2	1.281	1.304
3	0.862	0.858
4	1.308	1.310
5	0.937	0.929
6	0.603	0.603

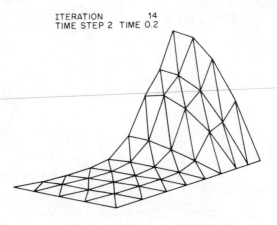

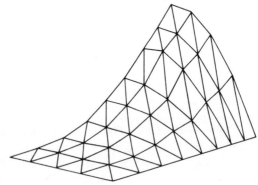

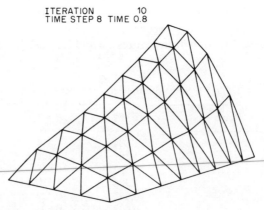

Figure 5.3

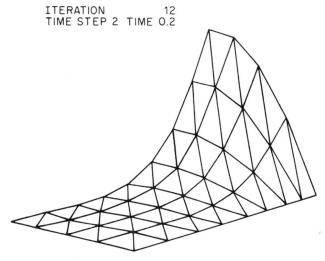

ITERATION   12
TIME STEP 2   TIME 0.2

Figure 5.4

oscillations gradually die out, whereas the unstable oscillation (with $\theta = 0.4$) in Figure 5.6 soon becomes much larger, and grows unboundedly.

5.3.3 Stability of time discretizations

The numerical example of the preceding section shows that, as with some of the examples in Chapter 4, the numerical solution can develop characteristics that mask the true nature of the analytic solution. The oscillations exhibited by the example can be explained by an analysis of the *stability* of the time discretization. The theta method applied to the semi-discrete system

$$B\dot{U} + GU = \mathbf{b}$$

can be written as

$$(B + \theta \Delta t G)U^{(n+1)} = (B - (1 - \theta)\Delta t G)U^{(n)} + \mathbf{b}^{(n)} \qquad (5.35a)$$

given an initial state $U^{(0)}$. If an alternative solution, denoted by $\bar{U}^{(n)}$, is based on the same discretization but starts from a different initial state $\bar{U}^{(0)}$, then the difference

$$\mathbf{E}^{(n)} = U^{(n)} - \bar{U}^{(n)}$$

satisfies

$$(B + \theta \Delta t G)\mathbf{E}^{(n+1)} = (B - (1 - \theta)\Delta t G)\mathbf{E}^{(n)} \qquad (5.35b)$$

given

$$\mathbf{E}^{(0)} = U^{(0)} - \bar{U}^{(0)}.$$

ITERATION   8
TIME STEP 2 TIME 0.2

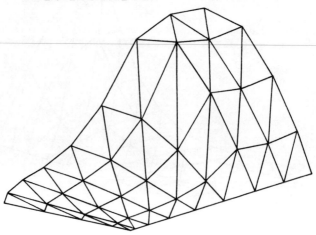

ITERATION   6
TIME STEP 4 TIME 0.4

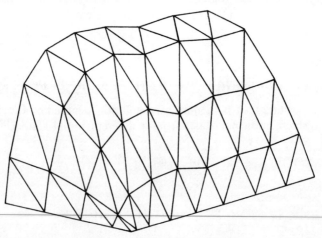

Figure 5.5

```
ITERATION          6
TIME STEP  4  TIME  0.4
```

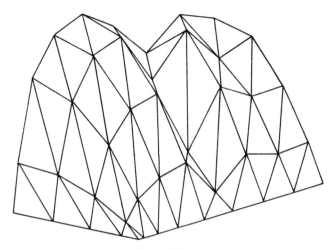

Figure 5.6

If the (error) $E^{(0)}$ is generated by round-off in the computer arithmetic, then $U^{(n)}$ and $\bar{U}^{(n)}$ can be viewed as two attempts to solve the same problem. A *stable* method is one in which the affect of the perturbation $E^{(0)}$ does not grow. If it is assumed that the analytic solution remains bounded then accurate numerical solutions will be obtained if *unstable methods are never used*.

The solution of (5.35b) is

$$E^{(n)} = M^n E^{(0)}$$

where

$$M = (B + \theta \Delta t G)^{-1}(B + (1 - \theta)\Delta t G). \tag{5.36}$$

The difference $E^{(n)}$ remains bounded as n tends to infinity if in some norm

$$\| M \| \leqslant 1 \tag{5.37}$$

and hence

$$\lim_{n \to \infty} M^n = 0.$$

In this case the effect of the error decays monotonically and (5.37) is a *sufficient* condition for stability. Alternatively, as for all matrix norms,

$$\rho^{n+1}(M) \leqslant \| M^{n+1} \| \leqslant \| M \|^{n+1}$$

where $\rho(.)$ is the *spectral radius* (i.e. the maximum modulus eigenvalue), it follows that

$$\rho(M) \leqslant 1$$

is a *necessary* condition for stability. Analysing the *amplification* matrix M is known as the *matrix method* of stability analysis.

It can be shown that for $\theta < \frac{1}{2}$ there is a critical value of Δt above which the matrix M in (5.36) has at least one eigenvalue greater than one in modulus. Thus for $\theta < \frac{1}{2}$ the method is *conditionally stable*; if $\theta \geqslant \frac{1}{2}$ the method is *unconditionally stable*.

The alternative *von Neumann* stability analysis assumes that the nodal values $\mathbf{U}^{(n)}$ and the errors $\mathbf{E}^{(n)}$ can be represented by a finite discrete Fourier series at each time level and such that each component is multiplied by a scalar *amplification factor* as the scheme proceeds to the next time level. Thus in one space dimension

$$E_k^{(n)} = \sum_j A_j e^{i\sigma_j x_k} e^{c_j \tau_n}$$

as the problem is linear, it is possible to consider a single component and assume that

$$E_k^{(n)} \approx e^{i\sigma x_k} e^{c\tau_n}.$$

Then, if the method is stable, the amplification factor

$$e^{c\Delta t} = \frac{E_k^{(n+1)}}{E_k^{(n)}}$$

must satisfy

$$|e^{c\Delta t}| \leqslant 1$$

for all c. The von Neumann analysis is easier to apply but it cannot take account of boundary effects.

As an example of the von Neumann analysis, consider the theta method applied to the piecewise linear approximation of

$$\frac{\partial u}{\partial t} = \frac{\partial^2 u}{\partial x^2}$$

given in Section 5.3.2. The nodal errors $E_k^{(n)}$ satisfy

$$\frac{h}{6}[E_{k+1}^{(n+1)} + 4E_k^{(n+1)} + E_{k-1}^{(n+1)}] + \frac{\theta}{h}\Delta t[-E_{k+1}^{(n+1)} + 2E_k^{(n+1)} - E_{k-1}^{(n+1)}]$$

$$= \frac{h}{6}[E_{k+1}^{(n)} + 4E_k^{(n)} + E_{k-1}^{(n)}] - (1-\theta)\frac{\Delta t}{h}[-E_k^{(n)} + 2E_k^{(n)} - E_{k-1}^{(n)}]. \quad (5.38)$$

Substituting

$$E_k^{(n)} = e^{i\sigma x_k} e^{c\tau_n}$$

then

$$E_{k\pm1}^{(n)} = e^{\pm i\sigma h} E_k^{(n)}$$

and with similar equations for $E_k^{(n+1)}$ etc. leads to

$$\left\{\frac{h}{6}[e^{i\sigma h} + 4 + e^{-i\sigma h}] + \frac{\theta}{h}\Delta t[-e^{i\sigma h} + 2 - e^{-i\sigma h}]\right\}e^{c\Delta t}E_k^{(n)}$$

$$= \left\{\frac{h}{6}[e^{i\sigma h} + 4 + e^{-i\sigma h}] - \frac{(1-\theta)}{h}\Delta t[-e^{i\sigma h} + 2 - e^{-i\sigma h}]\right\}E_k^{(n)}.$$

Using elementary trigonometric identities, this becomes

$$\left\{\frac{h}{3}\left[2\cos^2\left(\frac{\sigma h}{2}\right) + 1\right] + \frac{\theta}{h}\Delta t 4 \sin^2\left(\frac{\sigma h}{2}\right)\right\}e^{c\Delta t}$$

$$= \left\{\frac{h}{3}\left[2\cos^2\left(\frac{\sigma h}{2}\right) + 1\right] - \frac{(1-\theta)}{h}\Delta t 4 \sin^2\left(\frac{\sigma h}{2}\right)\right\}$$

or

$$e^{c\Delta t} = 1 - \frac{X}{1 + \theta X}$$

where

$$X = 12\frac{\Delta t}{h^2}\frac{\sin^2\left(\frac{1}{2}\sigma h\right)}{2\cos^2\left(\frac{1}{2}\sigma h\right) + 1}.$$

The stability condition requires $|e^{c\Delta t}| \leqslant 1$ for all c and hence for all σ. The worst case is $\sin\left(\frac{1}{2}\sigma h\right) = 1$ and this leads to the condition

$$\Delta t \leqslant \frac{h^2}{6(1 - 2\theta)}$$

if $\theta < \frac{1}{2}$. If $\theta \geqslant \frac{1}{2}$ the stability condition is satisfied for all σh and so the method is then unconditionally stable.

An analysis of the piecewise quadratic approximation is more complicated as the two distinct equations (one at integer nodes and one at mid-element nodes) have to be treated as a system of equations. Thus the von Neumann analysis leads to a 2×2 amplification matrix rather than a scalar amplification factor. The method is stable if both eigenvalues are less than one is modulus.

Exercise 5.5 Prove that a von Neumann analysis of the theta method applied to a piecewise bilinear solution of

$$\frac{\partial u}{\partial t} = \frac{\partial^2 u}{\partial x^2} + \frac{\partial^2 u}{\partial y^2}$$

leads to the stability condition

$$\Delta t \leqslant \frac{h^2}{12(1-2\theta)}$$

for $\theta < \frac{1}{2}$ with $\theta \geqslant \frac{1}{2}$ providing an unconditionally stable method.

Exercise 5.6 Verify that a von Neumann analysis of the piecewise quadratic approximation and the theta method for

$$\frac{\partial u}{\partial t} = \frac{\partial^2 u}{\partial x^2}$$

leads to the amplification matrix

$$\Gamma = [C + \Delta t \theta S]^{-1}[C - \Delta t(1 - \theta)S]$$

where

$$C = \frac{h}{10}\begin{bmatrix} 6 + 4\sin^2\left(\tfrac{1}{2}\sigma h\right) & 4\cos\left(\tfrac{1}{2}\sigma h\right) \\ 2\cos\left(\tfrac{1}{2}\sigma h\right) & 8 \end{bmatrix}$$

and

$$S = h^{-1}\begin{bmatrix} 12 + 4\cos^2\left(\tfrac{1}{2}\sigma h\right) & -16\cos\left(\tfrac{1}{2}\sigma h\right) \\ -8\cos\left(\tfrac{1}{2}\sigma h\right) & 8 \end{bmatrix}$$

i.e. the eigenvalues of Γ are

$$\lambda = \frac{1 - (1 - \theta)\Delta t\mu}{1 + \theta\Delta t\mu}$$

where μ is an eigenvalue of $M = C^{-1}S$.

5.3.4 Numerical stability of the CNG method

The numerical stability of (5.25) is governed by the behaviour of the amplification matrix

$$[B + \tfrac{1}{2}\Delta t G]^{-1}[B - \tfrac{1}{2}\Delta t G].$$

If the matrix $M = B^{-1}G$ has a complete set of eigenvectors and *positive* eigenvalues $\{\lambda_j\}, j = 1, 2, \ldots, N$, then the numerical stability is determined by the behaviour of the amplification factors

$$\mu_j = \frac{1 - \tfrac{1}{2}\Delta t \lambda_j}{1 + \tfrac{1}{2}\Delta t \lambda_j} \quad j = 1, 2, \ldots, N.$$

Clearly $|\mu_j| \leqslant 1$ if $\Delta t > 0$ but as N increases, $\lambda_N \to \infty$ and $\mu_N \to -1$. The solution of the semi-discrete equations (5.25) can be written as

$$U(t + \Delta t) = \exp(-\Delta t M)U(t) + g(t).$$

Thus the behaviour depends on the amplification matrix

$$\exp(-\Delta t M)$$

and hence on the amplification factors

$$v_j = e^{-\Delta t \lambda_j} \quad j = 1, 2, \ldots, N$$

such that $v_j \to 0$ as $\lambda_j \to 0$. Thus rapidly decaying components of the semi-discrete solution are replaced by oscillating and slowly decaying components of the discrete solution. Consequently errors introduced in the numerical calculation of $\mathbf{U}$ by (5.25) will only decay slowly and in an oscillatory manner. To prevent this a number of remedies have been suggested. The simplest is to restrict the time step Δt to ensure that $|\mu_N| < |\mu_1|$ (Lawson and Morris, 1978). Alternatively Linberg (1971) proposes a smoothing algorithm where $\mathbf{U}^{(1)}$ and $\mathbf{U}^{(2)}$ are calculated using (5.25). A revised $\mathbf{U}^{(1)}$ is then defined as

$$\mathbf{U}_{\text{rev}}^{(1)} = \tfrac{1}{4}(\mathbf{U}^{(0)} + 2\mathbf{U}^{(1)} + \mathbf{U}^{(2)})$$

and (5.25) used to calculate $\mathbf{U}_{\text{rev}}^{(n)}$, $n = 2, 3, \ldots$.

Various extrapolation procedures have been suggested by Gourlay and Morris (1980, 1981) in order to improve the accuracy of the time discretization. One possibility is to use two different theta methods: one with θ_1 defines $\mathbf{U}_1^{(n+1)}$, the other with $\theta_2 (\neq \theta_1)$ defines $\mathbf{U}_2^{(n+1)}$; then the linear combination

$$\mathbf{U}^{(n+1)} = \alpha \mathbf{U}_1^{(n+1)} + (1 - \alpha)\mathbf{U}_2^{(n+1)}$$

involves the three parameters θ_1, θ_2, and α that can be chosen to yield a more accurate solution. For the scheme to be second-order accurate in time it is necessary to have

$$\alpha\theta_1 + (1 - \alpha)\theta_2 = \tfrac{1}{2} \tag{5.39}$$

and for third-order

$$\alpha\theta_1^2 + (1 - \alpha)\theta_2^2 = \tfrac{1}{6}.$$

Alternatively, the amplification factor $\mu_j \to 0$ as $\lambda_j \to 0$ if in addition to (5.39) we have

$$\theta_1\theta_2 = (\theta_1 + \theta_2 - \tfrac{1}{2}).$$

Another possibility is to use one value of θ but to take a linear combination of the solution $\mathbf{U}_1^{(n+1)}$ after two time steps of length Δt and of the solution $\mathbf{U}_2^{(n+1)}$ after one time step of length $2\Delta t$. Then

$$\mathbf{U}^{(n+2)} = \beta\mathbf{U}_1^{(n+2)} + (1 - \beta)\mathbf{U}_2^{(n+2)}$$

where for second-order accuracy

$$(\beta - 2)(2\theta - 1) = 0$$

and third-order if $\theta = \tfrac{1}{2}$, $\beta = \tfrac{4}{3}$. If stability rather than third-order accuracy is required Lawson and Morris (1978) suggest $\theta = 0$, $\beta = 2$.

The extrapolation schemes involve the solution of more than one set of equations (5.26) at each time step. For example

$$[B + \theta_1 \Delta t G] U_1^{(n+1)} = \cdots$$

and

$$[B + \theta_2 \Delta t G] U_2^{(n+1)} = \cdots$$

at each time step or

$$[B + \theta \Delta t G] U_1^{(n+1)} = \cdots$$
$$[B + \theta \Delta t G] U_1^{(n+2)} = \cdots$$

and

$$[B + 2\theta \Delta t G] U_2^{(n+2)} = \cdots$$

for every two time steps. Thus two distinct matrices are used in each case.

Fairweather and Saylor (1983) suggest a *deferred correction* is preferable. Using the error expansions given by Fairweather and Johnson (1975) they show that, for example, with $\theta = 1$ the backward Euler method

$$B(U_1^{(n+1)} - U_1^{(n)}) + \Delta t G U_1^{(n+1)} = \Delta t b^{(n+1)} \tag{5.40}$$

a correction term $U_2^{(n+1)}$ can be computed to satisfy

$$B(U_2^{(n+1)} - U_2^{(n)}) + \Delta t G U_2^{(n+2)} = b_2^{(n+1)} \tag{5.41}$$

where

$$b_2^{(n+1)} = -\tfrac{1}{2} G(U_1^{(n+1)} - U_1^{(n)}) - \tfrac{1}{2}(b^{(n+1)} - b^{(n)}).$$

Then the solution is given by

$$U^{(n+1)} = U_1^{(n+1)} - \Delta t U_2^{(n+1)}.$$

This technique can be extended to higher-order accuracy, but always the equations have the same coefficient matrix, $[B + \Delta t G]$ in the case $\theta = 1$.

Alternatively a *defect correction method* (Fairweather and Saylor, 1983) can be used. This does not make use of the error expansions needed to define the right-hand side of (5.41). The values $\{U_1^{(n)}\}$ are computed from (5.40), are interpolated in time, and the residual or defect is computed and used as the right-hand side of an equation, analogous to (5.41), that defines the correction. Once again the same coefficient matrix $[B + \Delta t G]$ is used in both stages.

Hyperbolic equations (first and second order), parabolic equations with significant first derivatives, and nonlinear equations involving dispersion terms are amongst time-dependent partial differential equations which require careful discretization, first in space usually by a Petrov–Galerkin method, and then in time. It is not possible to give general guidelines on how to discretize these problems. In Chapter 7 numerical methods are given for a few particular examples of such problems.

5.3.5 One-step Galerkin methods: finite elements in time

An alternative approach is to discretize (5.22a) using a Galerkin approximation, that is, to derive an approximate solution which is of the form

$$V^{(n)}(t) = \sum_{j=0}^{S} V_j^{(n)} \varphi_j^{(n)}(t) \quad (n = 0, 1, \ldots) \tag{5.42}$$

in each subinterval $(\tau_n, \tau_n + \Delta t)$, where the coefficients $V_j^{(n)}(j = 1, \ldots, S)$ are determined by

$$\langle B\dot{V}^{(n)} + GV^{(n)}, \varphi_j^{(n)} \rangle_n = \langle \mathbf{b}, \varphi_j^{(n)} \rangle_n \quad (j = 1, \ldots, S; n = 0, 1, \ldots) \tag{5.43}$$

together with a continuity condition of the form

$$V^{(n)}(\tau_n^+) = V^{(n-1)}(\tau_n^-) \quad (n = 1, 2, \ldots). \tag{5.44}$$

The ith component of *vector* $\langle \mathbf{u}, v \rangle_n$ is

$$\int_{\tau_n}^{\tau_n + \Delta t} u_i(t)v(t)\, dt$$

where $\mathbf{u} = (u_1(t), \ldots, u_N(t))^T$.

Note that in (5.42), there are $S + 1$ basis functions, whereas there are only S equations in (5.43) and so the form of the difference approximation depends on the ordering of the basis functions.

We partition the subinterval $[\tau_n, \tau_n + \Delta t]$ such that

$$\tau_n = \tau_0^{(n)} < \tau_1^{(n)} < \cdots < \tau_S^{(n)} = \tau_n + \Delta t$$

with

$$\tau_j^{(n)} - \tau_{j-1}^{(n)} = \frac{\Delta t}{S} \quad (j = 1, \ldots, S).$$

We assume that $\varphi_j^{(n)}(j = 0, \ldots, S)$ form a basis for Lagrangian interpolation on $[\tau_n, \tau_n + \Delta t]$, that is,

$$\varphi_j^{(n)}(\tau_k^{(n)}) = \begin{cases} 1 & (j = k) \\ 0 & (j \neq k) \end{cases} \quad (n = 0, 1, \ldots)$$

and $\varphi_j^{(n)}(t)$ are polynomials of degree S on $[\tau_n, \tau_n + \Delta t]$. From (5.44) it follows that

$$V_0^{(n)} = V_S^{(n-1)} = U^{(n)} \approx U(\tau_n) \quad (n = 1, 2, \ldots)$$

and thus for $S \geq 2$ it is possible to eliminate $V_j^{(n)}(j = 1, \ldots, S - 1)$ from (5.43) to give a single equation for $U^{(n+1)}$ in terms of $U^{(n)}$ $(n = 0, 1, \ldots)$.

For example $(\mathbf{b} = 0)$:

(1) $S = 1$ (Comini *et al.*, 1974)

$$\left\{ B + \frac{2}{3}\Delta t G \right\} U^{(n+1)} = \left\{ B - \frac{\Delta t}{3} G \right\} U^{(n)}. \tag{5.45}$$

This is just the θ-method with $\theta = \frac{2}{3}$.

(2) $S = 2$

$$\left\{ I + \frac{3}{5}\Delta t M + \frac{3}{20}(\Delta t)^2 M^2 \right\} U^{(n+1)} = \left\{ I - \frac{2}{5}\Delta t M + \frac{1}{20}(\Delta t)^2 M^2 \right\} U^{(n)}$$

where $M = B^{-1}G$.

It has been shown (Hulme, 1972) that several well-known difference methods can be formulated as one-step Galerkin methods.

If we use Hermitian interpolation, different formulae are obtained. Another possible method of generating difference schemes is to replace the Galerkin approximation (5.43) by a least-squares approximation.

Exercise 5.7 Verify that if the basis functions are placed in reverse order, i.e.

$$\varphi_j^{(n)}(\tau_{S-k}^{(n)}) = \begin{cases} 1 & (k = j) \\ 0 & (k \neq j) \end{cases} \quad (n = 0, 1, \ldots)$$

and if $\mathbf{b} = 0$ and $S = 1$, we obtain the difference equation

$$\left\{ B + \frac{\Delta t}{3}G \right\} U^{(n+1)} = \left\{ B - \frac{2}{3}\Delta t G \right\} U^{(n)}.$$

Exercise 5.8 Different sets of difference equations can be obtained if the local approximation (5.42) is defined by

$$\langle B \dot{V}^{(n)} + G V^{(n)}, \psi_j^{(n)} \rangle_n = \langle \mathbf{b}, \psi_j^{(n)} \rangle_n \quad (j = 1, \ldots, S; n = 0, 1, \ldots)$$

in place of (5.43), where $\{\varphi_j^{(n)}\}$ and $\{\psi_j^{(n)}\}$ are dissimilar. Verify that with $\mathbf{b} = 0$,

$$\psi_1^{(n)}(t) = 1 \quad (n = 0, 1, \ldots; t \geq t_0)$$

and

$$\psi_2^{(n)}(t) = \frac{2}{\Delta t}(t - \tau_n) - 1 \quad (n = 0, 1, \ldots; \tau_{n-1} \leq t \leq \tau_n)$$

lead to:

(1) $S = 1$

$$\left\{ B + \frac{\Delta t}{2}G \right\} U^{(n+1)} = \left\{ B - \frac{\Delta t}{2}G \right\} U^{(n)} \quad \text{(i.e. CNG)}$$

(2) $S = 2$

$$\left\{ I + \frac{\Delta t}{2}M + \frac{(\Delta t)^2}{12}M^2 \right\} U^{(n+1)} = \left\{ I - \frac{\Delta t}{2}M + \frac{(\Delta t)^2}{12}M^2 \right\} U^{(n)}$$

where $M = B^{-1}G$.

Exercise 5.9 Verify that a least-squares approximation in place of (5.42) with $S = 1$ and $b = 0$ leads to the difference equation

$$\left\{\frac{1}{\Delta t}B^2 + \frac{1}{2}[BG + GB] + \frac{\Delta t}{3}G^2\right\}U^{(n+1)} = \left\{\frac{1}{\Delta t}B^2 + \frac{1}{2}[BG - GB] - \frac{\Delta t}{6}G^2\right\}U^{(n)}.$$

5.3.6 ADG methods for parabolic equations (Dendy and Fairweather, 1975; Fairweather, 1978)

In Chapter 4, ADG methods were introduced for elliptic problems. It is possible to use similar methods for parabolic problems, defined on rectangular regions. If tensor product basis functions are used, approximations defined by the *linear* equation

$$\left(\frac{\partial U}{\partial t}, V\right) + a(U, V) = (f, V) \quad (t > t_0)$$

lead (in two-dimensional problems) to an algebraic system that can be written as (see Section 4.5)

$$B_x \otimes B_y\left\{\frac{U^{(n+1)} - U^{(n)}}{\Delta t}\right\} + (G_x \otimes B_y + B_x \otimes G_y)\left\{\frac{U^{(n+1)} + U^{(n)}}{2}\right\} = b_{n+1/2}$$

$$(n = 1, 2, \ldots) \quad (5.46)$$

where $\otimes$ denotes tensor product. If the term $(\frac{1}{4}\Delta t)G_x \otimes G_y(U^{(n+1)} - U^{(n)})$ is added to the left-hand side, (5.46) can be replaced by

$$\left(B_x + \frac{\Delta t}{2}G_x\right) \otimes \left(B_y + \frac{\Delta t}{2}G_y\right)U^{(n+1)}$$

$$= \left(B_x - \frac{\Delta t}{2}G_x\right) \otimes \left(B_y - \frac{\Delta t}{2}G_x\right)U^{(n)} + b_{n-1/2}$$

which can be solved in two stages as for elliptic ADG methods (Section 4.5).

Several other methods have been proposed. In many cases attention has been restricted to linear problems, for example the application of general multistep methods (Zlámal, 1975) and Nørsett methods (Siemeniuch and Gladwell, 1974). Dupont *et al.* (1974) have constructed families of three-level difference schemes for solving both linear and nonlinear problems. Fairweather and Johnson (1975) have shown that it is possible to use local Richardson extrapolation based on such three-level schemes and also on certain two-level schemes proposed by Douglas and Dupont (1970). The latter authors (1975) have also considered the effect of interpolating the nonlinear coefficients, that is, interpolating functions such as $p(u)$ and $q(u)$ in (5.29).

Further details of semi-discrete Galerkin methods applied to time-dependent problems can be found in Chapter 7, where individual studies are carried out on problems from physics (solitons), engineering (convention–diffusion), biology (travelling pulses in axons), etc.

6

Convergence of Approximation

6.1 Introduction

In Chapter 4 the Galerkin approximation was introduced. If it is assumed that the solution u satisfies

$$a(u, v) = (f, v) \quad \text{(for all } v \in \mathcal{H})$$ (6.1a)

for a given $f \in \mathcal{H}$, it was shown how the approximation U satisfies an equation which can be written as

$$a(U, V) = (f, V) \quad \text{(for all } V \in K_N)$$ (6.1b)

where K_N is an N-dimensional subspace of the space of admissible functions $\mathcal{H}$. If it is assumed that the integrals are evaluated exactly, an analysis of the errors for piecewise smooth approximations of the form

$$U(\mathbf{x}) = \sum_{i=1}^{N} \alpha_i \varphi_i(\mathbf{x})$$

reduces to two distinct steps:

(1) A proof of the best or near-best properties of the approximation. A *best* approximation $U \in K_N$ in terms of the norm (1.11) is such that

$$\|u - U\| = \inf_{V \in K_N} \|u - V\|.$$ (6.2)

The norm in (6.2) is often very problem dependent and it is modified to show a Galerkin approximation given by (6.1) is *near best* in some *Sobolev norm*; that is

$$\|u - U\|_{r,R} \leqslant C \inf_{V \in K_N} \|u - V\|_{r,R}$$ (6.3)

170

for some $r, C > 0$.[†] This result usually follows from *Cea's lemma*.

(2) Prescribing an upper bound on the right-hand side of (6.3) by considering the particular case when $V \in K_N$ interpolates the solution. When K_N is a space of finite element approximations, there is usually an integer $k = k(K_N)$ and a constant $C = C(K_N)$ such that the error in interpolation is bounded as

$$\| u - V \|_{r,R} \leqslant Ch^{k+1-r} \| u \|_{k+1,R} \quad (r = 0, 1, \ldots, k) \tag{6.4}$$

provided $u \in \mathcal{H}^{(k+1)}(R)$.

The result is normally derived as a consequence of the *Bramble–Hilbert lemma*. Equations (6.3) and (6.4) can be combined to provide an estimate of the order of convergence of the Galerkin approximation U, as h tends to zero.

If, as usually happens, the integrals are evaluated numerically using a quadrature rule, then the approximate solution is no longer derived from (6.1) but from perturbed equations. The modified form can be written as

$$a_h(U_h, V) = (f, V)_h \quad \text{(for all } V \in K_N). \tag{6.5}$$

In such circumstances it is possible to derive an estimate of the order of convergence using (6.3) and (6.4) if a bound is available in the form

$$\| U - U_h \|_{r,R} \leqslant Ch^s$$

for some $s > 0$. Perturbations of the boundary also lead to modified equations in the form (6.5), as does the use of nonconforming elements, i.e. inadmissible functions such that $U_h \notin \mathcal{H}$. So it is clear that an analysis of problems defined by (6.5) is of considerable importance in the study of *practical* finite element methods.

6.1.1 The Bramble–Hilbert lemma

In this chapter the constants denoted by C may depend on the *region R* and the *space* $\mathcal{H}$ of admissible functions, but not on the particular function involved. It is assumed that the region R is an open bounded domain, with a (piecewise) smooth boundary ∂R, that $\bar{R} = R \cup \partial R$ and that, unless otherwise stated, $R \subset \mathbb{R}^2$. For sufficiently smooth functions—say in $C^k(\bar{R})$—use will occasionally be made of a *maximum norm* of the form

$$\| u \|_{(k)R} = \max_{|i| \leqslant k} \{ \| D^i u \| \mathcal{L}_{\infty(\bar{R})} \}$$

[†]In this chapter, C is used to denote a positive constant which will not be the same at each occurrence.

together with the corresponding semi-norm

$$|u|_{(k)R} = \max_{|i|=k} \{ \| D^i u \|_{\mathscr{L}_\infty(R)} \}.$$

The space $\mathring{\mathscr{H}}^{(k)}(R)$ is defined in Chapter 1 as

$$\mathring{\mathscr{H}}^{(k)}(R) = \{ u : u \in \mathscr{H}^{(k)}(R), D^\alpha u = 0 \text{ on } \partial R, |\alpha| < k \}.$$

Similarly for any finite-dimensional subspace $K_N \subset \mathscr{H}^{(k)}(R)$ define

$$\mathring{K}_N = K_N \cap \mathring{\mathscr{H}}^{(k)}(R)$$

and let $\bar{K}_N$ denote the complement of $\mathring{K}_N$ in K_N. For example for any 'finite element subspace' K_N, $\bar{K}_N$ is the subspace spanned by basis functions corresponding to the boundary nodes only; that is, basis functions which are zero at all the internal nodes.

In the interests of brevity, various additional restrictions on the region R will be omitted from the statements of the following lemmas. The restrictions are not mutually exclusive and are in general satisfied if the boundary is sufficiently smooth between the corners. The interested reader should consult the references cited for the details in each case.

Lemma 6.1 Sobolev's lemma (for example Agmon, 1965; or Yosida, 1971). *In $R \subset \mathbb{R}^m$, if $k > r + m/2$ then*

$$\| u \|_{(r)\bar{R}} \leqslant C \| u \|_{k,R}$$

Lemma 6.2 (Aubin, 1972, p. 180)

$$\| u \|_{0,R} \leqslant C |u|_{k,R} \quad \text{(for all } u \in \mathring{\mathscr{H}}^{(k)}(R)\text{)}.$$

Thus in particular

$$\| u \|_{0,R} \leqslant C |u|_{1,R} \quad \text{(for all } u \in \mathring{\mathscr{H}}^{(1)}(R)\text{)}.$$

If $\lambda(> 0)$ is the minimum eigenvalue of the Laplacian differential operator, subject to homogeneous Dirichlet boundary conditions, then $C = 1/\lambda$ (Courant and Hilbert, 1953, pp. 398–404). It follows from Lemma 6.2 that on $\mathring{\mathscr{H}}^{(1)}(R)$, $\|.\|_{1,R}$ and $|.|_{1,R}$ are equivalent norms.

The main result of this section is Theorem 6.1, the Bramble–Hilbert lemma. As it is of fundamental importance an attempt has been made to provide most of the proof, but in a book at this level it is impossible to give an explicit proof from first principles.

Lemma 6.3 Rellich's lemma   *The identity embedding of $\mathscr{H}^{(k)}(R)$ in $\mathscr{H}^{(k-1)}(R)$ is a compact mapping.*

This means that a sequence $\{u_s\}$ that is bounded in $\mathscr{H}^{(k)}(R)$, i.e.

$$\|u_s\|_{k,R} \leqslant 1 \quad \text{(say)}$$

has a subsequence $\{u_{s_j}\}$ that is convergent in $\mathscr{H}^{(k-1)}(R)$, i.e. there exists $u \in \mathscr{H}^{(k-1)}(R)$ such that

$$\lim \|u_{s_j} - u\|_{k-1,R} = 0.$$

Another interpretation is that sequences in $\mathscr{H}^{(k)}(R)$ may tend to limit functions that are themselves not sufficiently smooth to be in $\mathscr{H}^{(k)}(R)$ but they are in $\mathscr{H}^{(k-1)}(R)$.

A proof of Lemma 6.3 can be found *inter alia* in Nečas (1967), Adams (1975), Treves (1980), and Hutson and Pym (1980).

Lemma 6.4 (Nečas, 1967, p. 18)

$$\|u\|_{k,R}^2 \leqslant C \left\{ |u|_{k,R}^2 + \sum_{|i|<k} \left| \int \int D^i u \, dx \right|^2 \right\} \quad (\textit{for all } u \in \mathscr{H}^{(k)}(R)). \quad (6.6)$$

Proof The proof assumes that the inequality (6.6) does not hold (i.e. there exists no such C) and proceeds to show that this implies a contradiction.

If (6.6) does not hold then there exists a sequence $\{u_s\} \subset \mathscr{H}^{(k)}(R)$ such that the left-hand side remains finite while the right-hand side tends to zero, i.e.

$$\|u_s\|_{k,R} = 1 \quad \text{(say)}$$

and

$$\lim_{s \to \infty} \left\{ |u_s|_{k,R}^2 + \sum_{|i|<k} \left| \int \int_R D^i u_s \right|^2 \right\} = 0. \quad (6.7)$$

Thus

$$\lim D^i u_s = 0 \quad |i| = k.$$

From Lemma 6.3, it follows that as $\{u_s\}$ is a bounded sequence in $\mathscr{H}^{(k)}(R)$, there exists a convergent subsequence $\{u_{s_j}\}$ in $\mathscr{H}^{(k-1)}(R)$, and thus there exists $u \in \mathscr{H}^{(k-1)}(R)$ such that

$$\lim_{j \to \infty} u_{s_j} = u.$$

Thus

$$D^i u = \lim_{j \to \infty} D^i u_{s_j} = 0 \quad |i| = k$$

so $u \in \mathscr{H}^{(k)}(R)$ and is in fact a polynomial of degree at most $k-1$. Returning to (6.7), it follows that

$$\int \int_R D^i u = \lim_{j \to \infty} \int \int_R D^i u_{s_j} = 0 \quad |i| \leqslant k$$

and hence $u = 0$. Conversely,

$$\|u\|_{k,R} = \lim_{j \to \infty} \|u_{s_j}\|_{k,R} = 1$$

and $u \neq 0$. Thus we have the contradiction that proves the desired result.

Theorem 6.1 The Bramble–Hilbert lemma (Bramble and Hilbert, 1970; Ciarlet and Raviart 1972a) *Let $F \in \{\mathcal{H}^{(k+1)}(R)\}'$ be such that $F(p) = 0$ for all $p \in P_k$, the space of polynomials of degree at most k. Then there exists some constant $C = C(R)$ such that*

$$|F(u)| \leqslant C \|F\| \quad |u|_{k+1,R}$$

for any $u \in \mathcal{H}^{(k+1)}(R)$.

Proof It can be shown (exercise 6.1) that for any $u \in \mathcal{H}^{(k+1)}(R)$ there exists a polynomial $p = p(u) \in P_k$ such that

$$\iint_R D^i(u + p)\, \mathrm{d}\mathbf{x} = 0 \quad (|i| \leqslant k).$$

Thus from Lemma 6.4

$$\|u + p\|_{k+1,R}^2 \leqslant C |u + p|_{k+1,R}^2 = C |u|_{k+1,R}^2.$$

Then as the functional F is linear

$$F(u) = F(u + p)$$

and so combining all the results

$$|F(u)| \leqslant \|F\| \, \|u + p\|_{k+1,R} \leqslant C \|F\| \, |u|_{k+1,R}.$$

A frequent use of the Bramble–Hilbert lemma is in deriving bounds on bilinear forms. For example, let $\mathcal{H}$ be a Hilbert space and let F be a bounded bilinear form with arguments in $\mathcal{H}^{(k+1)}(R)$ and $\mathcal{H}$, that is, $F \in \mathcal{L}(\mathcal{H}^{(k+1)}(R) \times \mathcal{H}; \mathbb{R})$. Then if

$$F(u, v) = 0$$

for all $v \in \mathcal{H}$ when $u \in P_k$ and for some $v \in \mathcal{H}$, a functional $F_1 \in \{\mathcal{H}^{(k+1)}(R)\}'$ is defined by

$$F_1(u) = F(u, v) \quad \text{(for all } u \in \mathcal{H}^{(k+1)}(R))$$

the Bramble–Hilbert lemma leads to

$$|F_1(u)| \leqslant C \|F_1\| \, |u|_{k+1,R}.$$

In Section 1.2 (exercise 1.11), it is shown that for such a functional

$$\|F_1\| \leqslant \|F\| \, \|v\|_{\mathcal{H}}.$$

Hence, combining these results, it follows that

$$|F(u, v)| \leqslant C \|F\| \|v\|_{\mathscr{H}} |u|_{k+1, R}. \tag{6.8}$$

In Section 6.4 it is assumed that it is possible to define $\mathscr{H}^{(k)}(R)$ for non-integer values of k, as well as for integers. A detailed discussion of such a space and the significance of the *trace theorem* is beyond the scope of this book and the interested reader is referred to Aubin (1972), Lions and Magenes (1972), or Nečas (1967). A brief account of some of the more important properties and their uses can be found in Chapter 1 of Varga (1971).

Exercise 6.1 Prove by induction on k that for any $u \in \mathscr{H}^{(k+1)}(R)$ there exists a polynomial $p = p(u) \in P_k$ such that

$$\iint_R D^i(u + p) \, d\mathbf{x} = 0.$$

Exercise 6.2 Using Theorem 6.1, prove that if $F \in \mathscr{L}(\mathscr{H}^{(k+1)}(R) \times \mathscr{H}^{(r+1)}(R); \mathbb{R})$ is such that

$$F(u, v) = 0 \quad \begin{cases} \text{for all } u \in \mathscr{H}^{(k+1)}(R) \text{ if } v \in P_r \\ \text{for all } v \in \mathscr{H}^{(r+1)}(R) \text{ if } u \in P_k \end{cases}$$

then

$$|F(u, v)| \leqslant C \|F\| |u|_{k+1, R} |v|_{r+1, R}.$$

6.1.2 Regular transformations

It is usual to define basis functions on a standard element T_0, which could be either a unit square or a unit right triangle. A point transformation is then introduced to construct basis functions on an arbitrary element T (cf. Chapter 3). Thus it is natural to derive error estimates in terms of functions on a standard element, provided it is possible to transfer the bounds onto arbitrary elements.

In the case of curved elements, it is usual to consider the transformation in two parts by introducing an intermediate element T'. This intermediate element has the same vertices as the curved element T, but it has straight sides. Thus the mapping from T_0 onto T' is a linear transformation of the form

$$\mathbf{l} = \mathbf{F}_0(\mathbf{p})$$

such that

$$t = t_3 + (t_1 - t_3)p + (t_2 - t_3)q \quad (t = l, m)$$

where $\mathbf{p} = (p, q) \in T_0$ and $\mathbf{l} = (l, m) \in T'$. Then the mapping from T' onto the curved element T can be considered as a nonlinear perturbation of the linear transformation. If $\mathbf{x} = (x, y) \in T$, we write the complete transformation as

$$\mathbf{x} = \mathbf{F}(\mathbf{p}) = \mathbf{F}_0(\mathbf{p}) + \mathbf{F}_1(\mathbf{p})$$

where $\mathbf{F}_0$ is the linear transformation and $\mathbf{F}_1$ is the nonlinear perturbation term. The curved elements considered in Chapter 3 can all be considered in this way. Note that the *construction* of some of the elements uses a different sequence of transformations in which the intermediate element T' has curved sides and the same vertices as the standard triangle T_0.

In order that the bounds such as (6.8) can be used to derive estimates of the orders of convergence of individual finite element approximations it is often necessary to map $\mathscr{H}^{(r)}(T_0)$ onto $\mathscr{H}^{(r)}(T)$ and back again. Thus it is assumed that the transformation $\mathbf{F}$ is sufficiently differentiable to ensure that when $v \in \mathscr{H}^{(r)}(T)$, it follows that $v \circ \mathbf{F} \in \mathscr{H}^{(r)}(T_0)$, where the *composite operator* $v \circ \mathbf{F}$ is defined by

$$(v \circ \mathbf{F})(\mathbf{p}) = v(\mathbf{F}(\mathbf{p})) \quad \text{(for all } \mathbf{p} \in T_0).$$

To simplify the notation, we frequently write v in place of $v \circ \mathbf{F}$, if necessary writing $v(\mathbf{p})$ or $v(\mathbf{x})$ as appropriate, to avoid ambiguity. We also assume that the inverse transformation $\mathbf{F}^{-1}$ is sufficiently differentiable, thus when $v(\mathbf{p}) \in \mathscr{H}^{(r)}(T_0)$ it follows that $v(\mathbf{x}) \in \mathscr{H}^{(r)}(T)$.

Hypothesis 6.1 (The regularity hypothesis) If $v \in \mathscr{H}^{(r)}(T)$ and the diameter[†] of the element T is h, then it follows that there exist constants C_0, C_1 and C_2 such that

$$|v|_{r,T_0} \leqslant C_1 \left\{ \inf_{\mathbf{p} \in T_0} J(\mathbf{p}) \right\}^{-1/2} h^r \|v\|_{r,T} \tag{6.9a}$$

$$\|v\|_{r,T_0} \geqslant C_2 \left\{ \sup_{\mathbf{p} \in T_0} J(\mathbf{p}) \right\}^{-1/2} h^r \|v\|_{r,T} \tag{6.9b}$$

and

$$0 < \frac{1}{C_0} \leqslant \left\{ \frac{\sup J(\mathbf{p})}{\inf J(\mathbf{p})} \right\} \leqslant C_0 \tag{6.9c}$$

where J is the Jacobian of the transformation $\mathbf{x} = \mathbf{F}(\mathbf{p})$.

Note that this hypothesis implies that the Jacobian is positive for all $\mathbf{p} \in T_0$. In Chapter 3, it is shown that the Jacobian always satisfies this condition if the so-called forbidden elements are avoided.

When the linear transformation $\mathbf{x} = \mathbf{F}_0(\mathbf{p})$ is used, the Jacobian is constant and

$$J_0 = \det \begin{bmatrix} \dfrac{\partial x}{\partial p} & \dfrac{\partial x}{\partial q} \\[2mm] \dfrac{\partial y}{\partial p} & \dfrac{\partial y}{\partial q} \end{bmatrix} = \det \begin{bmatrix} 1 & x_1 & y_1 \\ 1 & x_2 & y_2 \\ 1 & x_3 & y_3 \end{bmatrix} = O(h^2)$$

and the bound (6.9c) is thus trivially satisfied.

[†]The diameter of a triangular element is the length of the longest side, the diameter of a quadrilateral element is the length of the longer diagonal.

For linear transformations, the hypothesis reduces to an easily verified scaling of the derivatives of a function u under the transformation T. In such a transformation

$$x, y \approx O(h)p + O(h)q + O(1)$$

$$\frac{\partial x}{\partial p}, \frac{\partial y}{\partial p}, \text{etc.} \approx O(h)$$

and thus

$$\frac{\partial u}{\partial p}, \frac{\partial u}{\partial q} \approx O(h)\frac{\partial u}{\partial x} + O(h)\frac{\partial u}{\partial y}. \tag{6.10}$$

Conversely

$$\frac{\partial u}{\partial x}, \frac{\partial u}{\partial y} \approx O\left(\frac{1}{h}\right)\frac{\partial u}{\partial p} + O\left(\frac{1}{h}\right)\frac{\partial u}{\partial q}. \tag{6.11}$$

Thus as

$$\iint_{T_0} u^2 J_0 \, dp \, dq = \iint_T u^2 \, dx \, dy$$

it follows that

$$\|u\|_{0,T_0} \approx O\left(\frac{1}{h}\right)\|u\|_{0,T} \tag{6.12a}$$

$$|u|_{1,T_0} \approx O(1)|u|_{1,T} \tag{6.12b}$$

and

$$|u|_{2,T_0} \approx O(h)|u|_{2,T}. \tag{6.12c}$$

The inequality (6.9b) is frequently replaced by

$$\sum_{r \leqslant M} h^r |u|_{r,T_0} \geqslant Ch^{M-1} \|u\|_{M,T}.$$

This is derived from the simple estimates after scaling by powers of h and summing.

The hypothesis has been verified for certain nonlinear transformations of the form

$$\mathbf{x} = \mathbf{F}_0(\mathbf{p}) + \mathbf{F}_1(\mathbf{p})$$

where $\mathbf{F}_1$ is a 'small' perturbation (see Exercises 6.5 and 6.6). If sufficient conditions are imposed on the perturbation term $\mathbf{F}_1$, it is possible to show that the Jacobian is

$$J(\mathbf{p}) = J_0 + J_1(\mathbf{p})$$

where J_1 is also 'small', and hence verify Hypothesis 6.1 (Ciarlet and Raviart, 1972c; Zlámal, 1974).

Exercise 6.3 Prove that it follows from (6.9b) that if $h < 1$ then

$$\|v\|_{r,T_0} \geqslant C\|v\|_{r,T} h^r \left\{ \sup_{p \in T_0} J(p) \right\}^{-1/2}.$$

Exercise 6.4 Prove that for any $v \in \mathscr{L}_2(T)$

$$\|Jv\|_{0,T_0} \leqslant \left\{ \sup_{p \in T_0} J(p) \right\}^{1/2} \|v\|_{0,T}.$$

Exercise 6.5 Prove that if $\mathbf{x} = \mathbf{F}_0(\mathbf{p})$, then there exist constants C_1 and C_2 such that

$$\iint_{T_0} \left(\frac{\partial v}{\partial t} \right)^2 d\mathbf{p} \leqslant C_1 h^2 \iint_T \left\{ \left(\frac{\partial v}{\partial x} \right)^2 + \left(\frac{\partial v}{\partial y} \right)^2 \right\} J_0^{-1} d\mathbf{x} \quad (t = p, q)$$

and

$$\iint_{T_0} \left\{ \left(\frac{\partial v}{\partial p} \right)^2 + \left(\frac{\partial v}{\partial q} \right)^2 \right\} d\mathbf{p} \geqslant C_2 h^2 \iint_T \left\{ \left(\frac{\partial v}{\partial x} \right)^2 + \left(\frac{\partial v}{\partial y} \right)^2 \right\} J_0^{-1} d\mathbf{x}.$$

Hence prove by induction that (6.9a) and (6.9b) are valid for $r = 1, 2, \ldots$ with $\|v\|_{r,T}$ replaced by $|v|_{r,T}$.

Exercise 6.6 (a) Verify that if P_5 and P_6 are the midpoints of the sides $P_2 P_3$ and $P_1 P_3$ respectively, then quadratic elements with one curved side (see Figure 6.1a) lead to the transformation

$$t = t_3 + (t_1 - t_3)p + (t_2 - t_3)q + \left(t_4 - \frac{t_1 + t_2}{2} \right) 4pq \quad (t = x, y).$$

(b) Verify that for such a transformation

$$J(\mathbf{p}) = J_0 + J_1(\mathbf{p})$$

where

$$J_1(\mathbf{p}) = 4\left(x_4 - \frac{x_1 + x_2}{2} \right)\{(y_2 - y_3)q - (y_1 - y_3)p\}$$

$$+ 4\left(y_4 - \frac{y_1 + y_2}{2} \right)\{(x_1 - x_3)p - (x_2 - x_3)q\}.$$

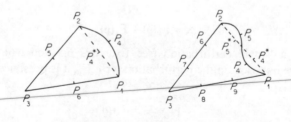

(a)   (b)

Figure 6.1

Find sufficient conditions to ensure that

$$0 < J_0 - Ch^3 \leqslant J \leqslant J_0 + Ch^3$$

and hence that $J = O(h^2)$. Compare your results with (6.42) in Section 6.3.2.

Zlámal (1973) has provided an alternative formulation for triangular elements with one curved side that leads to similar estimates.

Exercise 6.7 Verify that if P_6, P_7, P_8 and P_9 are the points of trisection of the sides $P_2 P_3$ and $P_1 P_3$ (see Figure 6.1b), then the cubic isoparametric elements lead to a transformation with nonlinear terms:

$$27pq(1 - p - q)\left(t_{10} - \frac{t_1 + t_2 + t_3}{3} \right)$$

$$+ \tfrac{9}{2}pq\left[(3p - 1)\left(t_4 - \frac{2t_1 + t_2}{3} \right) + (3q - 1)\left(t_5 - \frac{2t_2 + t_1}{3} \right) \right] \quad (t = x, y).$$

6.1.3 Complete approximations

The regularity hypothesis contains assumptions on the properties of the transformation from an arbitrary element to the standard element. When these assumptions are combined with the Bramble–Hilbert lemma it is possible to derive bounds such as (6.4) and hence determine the order of convergence of a finite element approximation. Thus the primary function of the Bramble–Hilbert lemma is to provide bounds on errors in interpolation. The lemma can be used equally well to bound the error in any form of approximation that is a projection onto a space of (piecewise) polynomials. In order to apply the lemma it is first necessary to make some assumptions concerning the types of function that make up the finite element approximation.

For any element T, we denote by $K_{[T]}$ the space defined by trial functions that can be nonzero in T; that is, $K_{[T]}$ is the restriction of the space of trial functions to the element T. We denote by $K_{[0]}$ the space of functions $v(\mathbf{p})$ such that $v(\mathbf{x}) \in K_{[T]}$. It is this space $K_{[0]}$ that has a critical significance in the analysis of finite element methods. The order of the method is governed by the maximum degree of polynomial (in $\mathbf{p}$) which, when interpolated by a function in $K_{[0]}$, leads to a zero interpolation error. In general this is equivalent to the maximum value of k such that $P_k \subset K_{[0]}$.

For example, if the approximation is in terms of piecewise cubic polynomials (Lagrange or Hermite) on a triangular mesh (Section 3.1), then the transformation $\mathbf{F}$ is linear and $K_{[0]}$, $K_{[T]} = P_3$. The eighteen-parameter C^1-quintics (3.14) leads to a linear transformation, $P_4 \subset K_{[0]}$, $K_{[T]} \subset P_5$ (strict inclusion), while if the biquadratic isoparametric approximation (3.37) is used, it follows that $P_2 \subset K_{[0]} \subset P_4$ (again strict inclusion) but that, as the transformation is nonlinear, $K_{[T]}$ is not a polynomial subspace.

For any element T, we introduce the projection Π_T onto $K_{[T]}$, that is, for any sufficiently differentiable function u, $\Pi_T u(\mathbf{x})$ interpolates $u(\mathbf{x})$ in T. Interpolation in this context means matching all the nodal parameters used in a finite element approximation. For functions defined on the standard element it is possible to define a mapping Π onto $K_{[0]}$ by

$$\Pi(v \circ \mathbf{F}) = (\Pi_T v) \circ \mathbf{F}.$$

Hypothesis 6.2 (The completeness hypothesis). If T is any element of diameter h, then there exists $k > 0$ such that $P_k \subset K_{[0]}$ and Π is a projection for which $\| I - \Pi \|$ is uniformly bounded for all h.

An example in which $\| I - \Pi \|$ is not uniformly bounded occurs using triangular elements when the normal derivative at a side point is a parameter but the function value and tangential derivative are not, and triangles tend to degenerate as the mesh is reduced, that is, there are elements in which the smallest angle tends to zero (Bramble and Zlámal, 1970). This example is mentioned again towards the end of Section 6.3.

We now have most of the building blocks necessary to construct an interpolation error bound. The details are delayed until Section 6.3 where Lemma 6.6 shows how to construct an interpolation error bound using the Bramble–Hilbert lemma on a reference element. The completeness hypothesis is invoked to guarantee that the operator norms remain bounded and then Theorem 6.5 illustrates how the regularity hypothesis is then used to transform the results onto physical elements at the same time introducing the powers of h expected from (6.4). The optimal properties of Galerkin approximations necessary to convert interpolation error bounds into finite element bounds are considered in Section 6.2.

The Bramble–Hilbert lemma combined with the regularity hypothesis is used primarily but not exclusively to estimate interpolation errors. It can be used equally well to derive bounds on the errors resulting from using numerical quadrature schemes based on finite element partitions of the region of integration.

Exercise 6.8 (i) Let $E_0(v)$ be the error in integrating $v \in \mathscr{H}^{(k+1)}(T_0)$ numerically over the standard element. If the quadrature rule is exact for polynomials of degree not exceeding k, verify that the Bramble–Hilbert lemma leads to the error bound

$$|E_0(v)| \leqslant C |v|_{k+1, T_0}.$$

(ii) Using the regularity hypothesis, verify that if the same rule is transformed onto an arbitrary element using a linear transformation, the error in integrating $v \in \mathscr{H}^{(k+1)}(T)$ can be written as $E_0(J_0 v)$, and that

$$|E_0(J_0 v)| \leqslant C h^{k+1} |v|_{k+1, T}.$$

Exercise 6.9 (i) Let $u \in \mathscr{H}^{(k+1)}(T_0)$ and $w \in P_r$, and denote by $E_0(uw)$ the error in integrating the product numerically over the standard element. By introducing the bilinear form

$E_1 \in \mathscr{L}(\mathscr{H}^{(k+1)}(T_0) \times \mathscr{L}_2(T_0);\mathbb{R})$ defined by

$$E_1(u, w) = E_0(uw)$$

prove that if the quadrature rule is exact for polynomials of degree not exceeding $r + k$, then there is a constant C such that

$$|E_0(uw)| \leqslant C \|w\|_{\mathscr{L}_2(T_0)} |u|_{k+1, T_0}.$$

(*Hint* The proof is analogous to the derivation of (6.4).)

(ii) Verify that if the quadrature rule is transformed onto an arbitrary element using a linear transformation, then the error in integrating the product of $u \in \mathscr{H}^{(k+1)}(T)$ and $w \in P_r$ numerically can be written as $E_0(J_0 uw)$ and that

$$|E_0(J_0 uw)| \leqslant Ch^{k+1} |u|_{k+1, T} \|w\|_{\mathscr{L}_0(T)}.$$

In Exercises 6.8 and 6.9 it is assumed that the transformations are linear as this ensures that when $w(\mathbf{x}) \in P_r$ then $J_0 w(\mathbf{p}) \in P_r$. If a nonlinear transformation is used as in Section 6.4.1, it is necessary to consider functions $w(\mathbf{x})$ such that $w(\mathbf{p}) J(\mathbf{p})$ is a polynomial. Further details of the transformation of quadrature rules onto arbitrary elements can be found in Section 6.4.1.

6.2 Convergence of Galerkin Approximations

6.2.1 Energy bounds

The first requirement is to establish the best or near-best (i.e. optimal) error bounds for conforming Galerkin approximation.

Theorem 6.2 Cea's lemma (Ciarlet, 1978) *Let $\mathscr{H}$ be a Hilbert space and a be a bounded $\mathscr{H}$-elliptic bilinear form, i.e.*

$$|a(u, v)| \leqslant \alpha \|u\| \|v\|$$

and

$$a(u, u) \geqslant \gamma \|u\|^2$$

then if $u \in \mathscr{H}$ and $U \in K_N \subset \mathscr{H}$ satisfy (6.1a) and (6.1b) respectively, the error is bounded as

$$\|u - U\| \leqslant \frac{\alpha}{\gamma} \inf_{V \in K_N} \|u - V\|. \tag{6.13}$$

if, in addition, the bilinear form is symmetric, then the approximation is best in terms of energy, i.e.

$$\|u - U\|_A = \inf_{V \in K_N} \|u - V\|_A. \tag{6.14}$$

Proof As $K_N \subset \mathscr{H}$ it follows from (6.1a) that

$$a(u, V) = (f, V) \quad V \in K_N.$$

Subtracting this from (6.1a) leads to

$$a(u - U, V) = 0 \quad V \in K_N. \tag{6.15}$$

Thus

$$a(u - U, u - U) = a(u - U, u - V) + a(u - U, V - U),$$

but

$$a(u - U, V - U) = 0$$

hence applying the positivity on the left and the continuity on the right leads to

$$\gamma \|u - U\|^2 \leqslant a(u - U, u - U) = a(u - U, u - V) \leqslant \alpha \|u - U\| \|u - V\|$$

i.e.

$$\|u - U\| \leqslant \frac{\alpha}{\gamma} \|u - V\|.$$

The second result, (6.14b), follows directly from (6.15), since as the bilinear form is symmetric this can be written in terms of the energy inner product as

$$(u - U, V)_A = 0$$

and it follows directly form Theorem 1.2 that U is the best approximation in terms of the energy norm $\|\cdot\|_A$.

Corollary (i) *The Galerkin equations are equivalent to the variational principle*

$$\min_{U_h \in K_N} \|u - U_h\|_A^2. \tag{6.16}$$

As an example of the results that can be derived from Theorem 6.2, consider a Ritz approximation to the solution of

$$\frac{\partial}{\partial x}\left(d_1(x, y)\frac{\partial u}{\partial x}\right) + \frac{\partial}{\partial y}\left(d_2(x, y)\frac{\partial u}{\partial y}\right) + f(x, y) = 0 \quad ((x, y) \in R) \tag{6.17}$$

subject to

$$u(x, y) = 0 \quad ((x, y) \in \partial R) \tag{6.18}$$

where there exist constants δ and Δ such that

$$0 < \delta \leqslant d_1(x, y), d_2(x, y) \leqslant \Delta \quad ((x, y) \in R).$$

In this application

$$l(v) = (f, v)$$

and

$$a(u, v) = \iint_R \left\{ d_1 \frac{\partial u}{\partial x} \frac{\partial v}{\partial x} + d_2 \frac{\partial u}{\partial y} \frac{\partial v}{\partial y} \right\} dx\, dy.$$

Since

$$|a(u,v)| \leqslant \Delta \int\int_R \left\{ \left| \frac{\partial u}{\partial x} \frac{\partial v}{\partial x} \right| + \left| \frac{\partial u}{\partial y} \frac{\partial v}{\partial y} \right| \right\} dx\,dy$$

it is bounded and since

$$a(u,u) \geqslant \delta |u|_{1,R}$$

it follows from Lemma 6.2 that it is $\overset{\circ}{\mathscr{H}}{}^{(1)}(R)$-elliptic.

Corollary (ii) *If u is the solution of* (6.17) *and* (6.18), *then the Galerkin approximation $U \in \overset{\circ}{K}_N \subset \overset{\circ}{\mathscr{H}}{}^{(1)}(R)$ is such that*

$$\|u - U\|_{1,R} \leqslant C \inf_{\tilde{u} \in \overset{\circ}{K}_N} \|u - \tilde{u}\|_{1,R}.$$

If it is necessary to approximate the boundary conditions in terms of basis functions that take nonzero values on the boundary, it is still possible to derive error bounds. If

$$U = U_0 + \bar{U}$$

where $U_0 \in \overset{\circ}{K}_N \subset \overset{\circ}{\mathscr{H}}{}^{(1)}(R)$ then $\bar{U}$ contains *only* basis functions that can take nonzero values on the boundary and is completely determined by the boundary data. It follows from the definitions in Section 6.1 that $\bar{U} \in \bar{K}_N$.

Lemma 6.5 *If u is the solution of* (6.17), *subject to $u = g$ on ∂R, and $\bar{U} \in \bar{K}_N$ is chosen such that $\bar{U}(x,y)$ $((x,y) \in \partial R)$ is a fixed approximation to g, then the finite element approximation $U = U_0 + \bar{U}$ such that $U_0 \in \overset{\circ}{K}_N \subset \overset{\circ}{\mathscr{H}}{}^{(1)}(R)$ is such that*

$$|u - U|_{1,R} \leqslant C|u - (\tilde{u} + \bar{U})|_{1,R} \quad (for\ all\ \tilde{u} \in K_N).$$

Proof Since

$$u - (\tilde{u} + \bar{U}) = u - U + U_0 - \tilde{u}$$

and $U_0 - \tilde{u} \in \overset{\circ}{K}_N$ the result follows directly from Lemma 6.4 and the definitions of u and U.

It is also possible to use Lemma 6.5 to derive bounds in terms of $\mathscr{H}^{(1)}(R)$ if we assume that there exists a smooth continuation of g into R. At least one such continuation exists, namely the solution u itself.

Theorem 6.3 (Fairweather, 1978, p. 45) *Let u be the solution of* (6.13), *subject to $u = g$ on ∂R, and let $w \in \mathscr{H}^{(1)}(R)$ be any smooth continuation of g into R. Then the finite element approximation U, which can be written as $U = U_1 + W$ where $W \in K_N$, is an approximation to w and $U_1 \in \overset{\circ}{K}_N$, is such that*

$$\|u - (U_1 + W)\|_{1,R} \leqslant C\{\|u - (\tilde{u} + W)\|_{1,R} + \|w - W\|_{1,R}\} \tag{6.19}$$

for any $\tilde{u} \in \overset{\circ}{K}_N$.

Note that if, for example, the boundary condition is interpolated, the right-hand side of (6.19) consists of the error in approximating $u - w \in \mathscr{H}^{(1)}(R)$ together with the error interpolating the function w, which is nonzero on the boundary.

Exercise 6.10 Prove that if (i) u is the solution of (6.17) subject to $u = g$ on ∂R and (ii) $w \in \mathscr{H}^{(1)}(R)$ is *any* function such that $w = g$ on ∂R then the Galerkin approximation $U + w$ with $U \in \overset{\circ}{K}_N \subset \overset{\circ}{\mathscr{H}}^{(1)}(R)$ is such that

$$\| u - (U + w) \|_{1,R} \leqslant C \inf_{\tilde{u} \in \overset{\circ}{K}_N} \| u - (\tilde{u} + w) \|_{1,R}.$$

Exercise 6.11 Prove that, if u is a solution of (6.17), subject to $\partial u / \partial n = g$ on ∂R, then there exists a Galerkin approximation $U \in K_N \subset \mathscr{H}^{(1)}(R)$ such that

$$\| u - U \|_{1,R} \leqslant C \inf_{\tilde{u} \in K_N} \| u - \tilde{u} \|_{1,R}.$$

Note that it is necessary to assume a *compatibility condition* on f and g if the solution is to exist. For example, if $d_1 = d_2 = 1$ in (6.17), then we assume that

$$\iint_R f \, dx \, dy = - \int_{\partial R} g \, ds$$

and as the solution is only unique to within an additive constant, it is possible to choose a constant such that

$$\iint_R u \, dx \, dy = 0$$

(Nečas, 1967, p. 256). Hence it is possible to apply Lemma 6.4 with $k = 1$, to obtain the result.

6.2.2 Alternative norms

The estimates provided by Cea's lemma, together with the Bramble–Hilbert lemma, indicate optimality in energy or an equivalent Sobolev norm. These Sobolev norms involve derivatives and thus the error estimates provide bounds on the derivatives of the error as well as the error itself. It may be the case, as in plane stress or plane strain problems, that the derivatives are the important physical quantities and such bounds are very useful. Alternatively, if the derivatives are not required explicitly, a bound on the $\mathscr{L}_2$ norm might be more appropriate. Since for any u,

$$\| u \|_{0,R} \leqslant \| u \|_{1,R}$$

it is easy to obtain a bound on the $\mathscr{L}_2$ error, but as the error is dominated by the error in the derivatives, this simple bound will not be optimal.

The fundamental result used to modify optimal error bounds in one norm to provide optimal bounds in a weaker (i.e. more general) norm is the *Aubin–Nitsche lemma*. Lemms 6.3 introduced the notion of an embedding, that is surrounding a

Hilbert space by a slightly bigger Hilbert space of less smooth functions. A *dense embedding* is such that the closure of the smaller space fills the larger space, e.g. for any function $u \in \mathscr{L}_2(R)$ it is possible to construct a sequence $\{u_s\} \subset \mathscr{H}^{(1)}(R)$ (not necessarily bounded in $\|\cdot\|_1$), such that

$$u = \lim_{s \to \infty} u_s.$$

If H_1 is a Hilbert space, densely embedded in the Hilbert space H_0, we denote this by $H_1 \subset H_0$ and $\bar{H}_1 = H_0$.

Theorem 6.4 Aubin–Nitsche lemma (Ciarlet, 1978; Showalter, 1977) *Let $H_1 \subset H_0$ and $\bar{H}_1 = H_0$ with norms $\|.\|_1$ and $\|.\|_0$ respectively, and assume in addition that $u \in H_1$ satisfies*

$$a(u, v) = F(v) \quad \text{for all } v \in H_1 \tag{6.20}$$

and

$$a(U, V) = F(V) \quad \text{for all } V \in K_N \subset H_1 \tag{6.21}$$

where a is an H_1-elliptic bounded bilinear form and $F \in H_1'$. Then if

$$e = \frac{u - U}{\|u - U\|_0} \tag{6.22}$$

and

$$a(v, w) = (e, v)_0 \quad \text{for all } v \in H_1 \tag{6.23}$$

and

$$a(V, W) = (e, V)_0 \quad \text{for all } V \in K_N \subset H_1 \tag{6.24}$$

then

$$\|u - U\|_0 \leqslant C \|u - U\|_1 \|w - W\|_1. \tag{6.25}$$

Thus the error in $\|.\|_0$ is bounded in terms of the error in $\|.\|_1$, with the additional term providing the extra powers of h. In applications in this book, we identify H_1 with $\mathscr{H}^{(r)}(R)$ and H_0 with $\mathscr{L}_2(R)$.

Proof From (6.16) and (6.17),

$$a(u - U, W) = 0 \tag{6.26}$$

and (from (6.22))

$$
\begin{aligned}
0 \leqslant \|u - U\|_0 &= (e, u - U)_0 \\
\text{(from (6.23))} \quad &= a(u - U, w) \\
\text{(from (6.26))} \quad &= a(u - U, w) - a(u - U, W) \\
&= a(u - U, w - W) \\
\text{(from continuity)} \quad &\leqslant C \|u - U\|_1 \|w - W\|_1.
\end{aligned}
$$

Given the (lack of) smoothness in the error function e, it is shown in the next section how the last term in (6.25) provides just the correct modification to yield the optimal estimate. This analysis is appropriate for $\mathscr{L}_2$ bounds; the highly technical results for $\mathscr{L}_\infty$ bounds are summarized in Section 6.3.3.

6.2.3 Boundary approximation and quadrature

Computing a finite element approximation with interpolated boundary conditions—as above—is only one of the major *variational crimes* (Strang, 1972) invariably committed when solving practical problems. The others are (i) altering the position of the boundary, (ii) using numerical integration (quadrature) for the inner products, and (iii) using nonconforming elements. If any of these techniques are used then it follows that the approximate solution is no longer in K_N nor satisfies

$$a(U, V) = (f, V) \quad \text{(for all } V \in K_N\text{)}. \tag{6.27}$$

It is instead $U_h \in K_h$ and satisfies

$$a_h(U_h, V_h) = (f, V_h)_h \quad \text{(for all } V_h \in K_h\text{)} \tag{6.28}$$

where both sides of (6.28) and the form of the new M-dimensional space K_h, which may contain functions that are not admissible for the classical variational methods outlined in Chapter 4, depend on the nature of the variational crimes committed. It is usually the case that $M \geqslant N$ and that $K_h \supseteq K_N$. If $K_h = K_N$, as when the use of quadrature is the only deviation from the classical form of the variational method, the solutions of both (6.27) and (6.28) are linear combinations of $\varphi_i(x)(i = 1, \ldots, N)$. The coefficients of the solution to (6.27) are given by

$$G\alpha = \mathbf{b} \tag{6.29}$$

where the *stiffness matrix* $G = \{a(\varphi_i, \varphi_j)\}(i = 1, \ldots, N)$ and $\mathbf{b} = \{(f, \varphi_i)\}$. The perturbed problem (6.28) leads to

$$G_h \alpha_h = \mathbf{b}_h \tag{6.30}$$

where $G_h = \{a_h(\varphi_i, \varphi_j)\}$ and $\mathbf{b}_h = \{(f, \varphi_i)_h\}$. It is possible to compare the individual elements of G_h and $\mathbf{b}_h$ with the corresponding elements of G and $\mathbf{b}$, then derive bounds on $\|U - U_h\|$ by a standard perturbation analysis of linear algebra. Unfortunately, this method is known to lead to a poor upper bound on $\|U - U_h\|$ in most cases (Fix, 1972). If it is assumed that the perturbed form is $\mathscr{H}$-elliptic and bounded then not only is the existence of a unique solution guaranteed by Theorem 6.2, but it is also possible to estimate the size of the perturbation.

These results are fundamental to the analysis of Sections 6.4.1–6.4.5 but there is not room here to provide detailed proofs, so they are presented as a sequence of exercises 6.12–6.14. Note that the estimate (6.31) can only be used when U_h

is in some sense a perturbation of a classical Galerkin approximation. That is, if $K_h = K_N$ or $K_h = K_N \oplus \{\varphi_{N+1}, \ldots, \varphi_M\}$, where the additional $\varphi_i(x)$ $(i = N+1, \ldots, M)$ have special properties that exclude them as admissible functions for a classical approximation; they could be nonconforming for example. Alternatively the estimate (6.32) is of use when the approximation differs significantly from any classical approximation such as when nonconforming elements alone are used.

The same form of perturbation analysis can be used when (6.28) represents a system of *variational difference equations* (for example, Dem'janovič, 1964).

Exercise 6.12 Verify that if a_h is an $\mathcal{H}$-elliptic bilinear form then

$$a_h(U - U_h, W_h) = (a_h - a)(U, W_h) + (f, W_h) - (f, W_h)_h$$

for any $W_h \in K_h$. Hence prove that

$$\|U - U_h\| \leqslant C \sup_{W_h \in K_N} \left\{ \frac{|(a_h - a)(U, W_h)| + |(f, W_h) - (f, W_h)_h|}{\|W_h\|} \right\} \tag{6.31}$$

and construct a similar bound for $\|u - U_h\|_{\mathcal{H}}$.

Exercise 6.13 Verify that if a_h is a bounded bilinear form then

$$|a_h(U_h - V_h, W_h)| \leqslant \alpha \|u - V_h\| \|W_h\| + |(f, W_h)_h - a_h(u, W_h)|$$

for any $V_h, W_h \in K_h$. Hence prove that if a_h is also $\mathcal{H}$-elliptic

$$\|U_h - V_h\| \leqslant \frac{\alpha}{\gamma} \|u - V_h\| + \sup_{W_h \in K_h} \left\{ \frac{1}{\gamma} \frac{|(f, W_h)_h - a_h(u, W_h)|}{\|W_h\|} \right\}$$

for any $V_h \in K_h$, and that there exists $C > 0$ such that

$$\|u - U_h\| \leqslant C \left\{ \inf_{V_h \in K_h} \|u - V_h\| + \sup_{W_h \in K_h} \left\{ \frac{|(f, W_h)_h - a_h(u, W_h)|}{\|W_h\|} \right\} \right\}. \tag{6.32}$$

Exercise 6.14 Prove that if a_h and a are bilinear forms such that a_h is K_h-elliptic and a is bounded then

$$\|u - U_h\| \leqslant C \left\{ \inf_{V_h \in K_h} \left[\|u - V_h\| + \sup_{W_h \in K_h} \left\{ \frac{|(a_h - a)(V_h, W_h)|}{\|W_h\|} \right\} \right] \right.$$
$$\left. + \sup_{W_h \in K_h} \left\{ \frac{|(f, W_h)_h - (f, W_h)|}{\|W_h\|} \right\} \right\}. \tag{6.33}$$

6.3 Approximation Errors

As is shown in the previous section, Galerkin approximations to elliptic problems are invariably *near best* in some norm. In particular, from the corollary to Cea's lemma applied to second-order problems, it follows that

$$\|u - U\|_{1,R} \leqslant C \inf_{\tilde{u} \in K_N} \|u - \tilde{u}\|_{1,R}. \tag{6.34}$$

Bounds on the error have also been derived in terms of other norms, for example a bound on the $\mathscr{L}_2(R)$ norm for second-order problems can be derived from the Aubin–Nitsche lemma as

$$\|u - U\|_{0,R} \leqslant C\|u - U\|_{1,R}\|w - W\|_{1,R} \qquad (6.35)$$

where w and W correspond to the subsidiary problem with the error $(u - U)$ as the right-hand side.

6.3.1 Interpolation error bounds and Galerkin error bounds

Let T_0 be the *standard element*. Then most of the convergence results can be derived from the following lemma concerning polynomial approximation on T_0.

Lemma 6.6 Let $\Pi \in \mathscr{L}(\mathscr{H}^{(k+1)}(T_0); \mathscr{H}^{(r)}(T_0))\ (k \geqslant r)$ *be a projection onto* $K_{[0]}$ *where* $\mathscr{H}^{(r)}(T_0) \supset K_{[0]} \supset P_k$. *Then*

$$\|v - \Pi v\|_{r,T_0} \leqslant C\|I - \Pi\|\,|v|_{k+1,T_0}$$

for all $v \in \mathscr{H}^{(k+1)}(T_0)$.

That is, if the interpolation is exact for polynomials of degree not exceeding k, the error in interpolation can be expressed in terms of the $(k+1)$th derivatives of the function being interpolated. The operators Π and $I - \Pi$ are defined from $\mathscr{H}^{(k+1)}(T_0)$ onto $\mathscr{H}^{(r)}(T_0)$ because when this result is applied later, r, depends on the order of the differential equation and k depends on the properties of the trial functions. The value of $\|I - \Pi\|$ depends on the values of r and k, but it is assumed to be uniformly bounded and the precise value is not, in general, important.

Proof. For any $G \in \{\mathscr{H}^{(r)}(T_0)\}'$ define $F \in \{\mathscr{H}^{(k+1)}(T_0)\}'$ such that for any $v \in \mathscr{H}^{(k+1)}(T_0)$

$$F(v) = G([I - \Pi]v).$$

Then by applying the Bramble–Hilbert lemma to F, we have that

$$|G([I - \Pi]v)| \leqslant C\|F\|\,|v|_{k+1,T_0}$$

where

$$\|F\| = \sup_{w \in \mathscr{H}^{(k+1)}(T_0)} \left\{ \frac{|G([I - \Pi]w)|}{\|w\|_{k+1,T_0}} \right\}.$$

It follows from duality that for any $u \in \mathscr{H}^{(r)}(T_0)$

$$\|u\|_{r,T_0} = \sup_G \left\{ \frac{|G(u)|}{\|G\|} \right\}.$$

Thus in particular

$$\|(I - \Pi)v\|_{r,T_0} = \sup_G \left\{ \frac{|G[I - \Pi]v)|}{\|G\|} \right\}.$$

Then, combining these results, it follows that

$$\|(I - \Pi)v\|_{r,T_0} \leqslant C|v|_{k+1,T_0} \sup_G \left\{ \frac{\|F\|}{\|G\|} \right\}$$

$$\leqslant C|v|_{k+1,T_0} \sup_G \left\{ \sup_{w \in \mathscr{H}^{(k+1)}(T_0)} \left\{ \frac{|G([I - \Pi]w)|}{\|w\|_{k+1,T_0}} \right\} \frac{1}{\|G\|} \right\}$$

but

$$|G([I - \Pi]w)| \leqslant \|G\| \, \|I - \Pi\| \, \|w\|_{k+1,T_0}$$

and so the result follows immediately.

This result can be combined with the regularity hypothesis to provide an interpolation error bound in the region R.

Theorem 6.5 Assume that the completeness hypothesis is valid for some $k > 0$ and that the regularity hypothesis is valid for all $r \leqslant k$. Then if $u \in \mathscr{H}^{(k+1)}(R)$ and $\tilde{u}$ interpolates u, it follows that

$$\|u - \tilde{u}\|_{r,R} \leqslant Ch^{k+1-r}\|u\|_{k+1,R} \tag{6.36}$$

where h is a bound on the diameters of the elements of the partition of R.

Proof. If we assume that the region R is partitioned into elements $T_j (j = 1, \ldots, S)$ then

$$\|u - \tilde{u}\|_{r,R}^2 = \sum_{j=1}^{S} \|u - \Pi_{T_j} u\|_{r,T_j}^2.$$

Considering a typical element T, it follows from Lemma 6.6 that,

$$\|u - \Pi u\|_{r,T_0} \leqslant C\|I - \Pi\| \, |u|_{k+1,T_0}.$$

Then by Hypothesis 6.1, applying (6.9a) on the right and (6.9b) on the left it follows that

$$\|u - \Pi_T u\|_{r,T} \leqslant Ch^{k+1-r}\|I - \Pi\| \, \|u\|_{k+1,T}.$$

The bound (6.9c) is also required if C is bounded. If in addition Hypothesis 6.2 is assumed, then there exists a uniform bound on all the operator norms $\|I - \Pi\|$. Summing over all elements leads to the desired result.

One particular situation in which this result can be applied is that in which the transformation T_0 to T is linear and interpolation by finite element approximating functions is exact for $u \in P_k$. Then in particular $J = $ constant and (6.9a) holds with $|.|_r$ in place of $\|.\|_r$. Thus

$$\|u - \tilde{u}\|_{1,R} \leqslant Ch^k |u|_{k+1,R}$$

where $\tilde{u}$ is the interpolating function. This is the type of bound required to estimate the order of the finite element approximation for second-order problems using the near-best inequalities of Section 6.2.1. Note that if we consider a region made up of one element we have an error bound for classical Lagrange interpolation; this will be used later, in Section 6.4.2.

Corollary (i) *The Lagrangian (or Hermitian) elements of Section 3.1 are such that the transformation from T_0 to T is linear, the basis functions $\varphi \in P_k$, interpolation is exact for $u \in P_k$, and $K_{[0]} = K_{[T]} = P_k$. Thus if the region R is a polygon with the boundary conditions matched exactly and the integrals evaluated analytically, the error in such a finite element approximation to a second-order problem is such that*

$$\|u - U\|_{1,R} \leqslant Ch^k |u|_{k+1,R}. \tag{6.37}$$

An equivalent result holds for problems on rectangles with a rectangular grid; again it is the degree of polynomials that can be interpolated exactly that governs the exponent of h in the error bound (6.37).

If the solution is not sufficiently smooth $|u|_{k+1,R}$ may not exist and it will not be possible to derive a bound such as (6.37), even though $K_{[0]} \supset P_k$. In such cases, it is necessary to use the exponent

$$k^* = \max \{s : u \in \mathcal{H}^{(s)}(R)\}.$$

Since $k^* \leqslant k$, the only modification of Theorem 6.5 that is necessary is to replace k by k^* wherever it appears; the analysis is unaltered.

This reduction in the order of the error is typical of problems with re-entrant corners. Such problems also cause problems for the $\mathcal{L}_\infty$ error analysis discussed later. The lack of smoothness is important when implementing (6.36) in the $\mathcal{L}_2$ error bound (6.35). As detailed in the previous section, w and W are the solutions of

$$a(v, w) = (e, v) \quad \text{for all } v \in H_1 \tag{6.38}$$

and

$$a(V, W) = (e, V) \quad \text{for all } V \in K_N \subset H_1$$

respectively. The function e is the $\mathcal{L}_2$ error in the Galerkin approximation, scaled so that

$$\|e\|_{0,R} = 1. \tag{6.39}$$

Lemma 6.7 *If a is an $\mathcal{H}^{(r)}$-elliptic bilinear form and $(,)$ is the $\mathcal{L}_2$ inner product,* *then the solution of* (6.38) *satisfies*

$$\|w\|_{2r,R} \leqslant C\|e\|_0.$$

Note on proof If a is $\mathscr{H}^{(r)}$-elliptic, then (6.38) is the weak form of a differential equation of order $2r$, so that that $2r$th derivatives of the solution w have the same smoothness as the right-hand side e.

Corollary (ii) *For w defined by* (6.38), *the inequality* (6.36) *becomes*

$$\|w - W\|_{r,R} \leqslant Ch^r \tag{6.40}$$

and so for second-order problems $(r = 1)$, *subject to the same conditions as Corollary* (i),

$$\|u - U\|_{0,R} \leqslant Ch^{k+1}|u|_{k+1,R}. \tag{6.41}$$

Interpolation error bounds in terms of the maximum semi-norm $|.|_{(k)\bar{R}}$ rather than the Sobolev semi-norm $|.|_{k,R}$ were first derived by Zlámal and later by Ciarlet and Raviart and by Zeníšek (Ciarlet and Raviart, 1972a, and references therein).

Exercise 6.15 Verify that eliminating the internal parameters from finite-element approximations such as cubic Lagrangian (or Hermitian) elements, or eliminating the normal derivatives at the side midpoints of the 21-parameter C^1-quintics, leads to a reduction of one in the exponent of h in (6.37).

Exercise 6.16 Verify that subparametric biquadratic and bicubic approximation, using a bilinear transformation from an arbitrary quadrilateral element onto the unit square, can interpolate exactly quadratic and cubic polynomials respectively, in x and y. Show that if R is a polygon, the boundary conditions are matched exactly, and the integrals are evaluated analytically, then the error bound (6.37) is valid for $k = 2$ and 3 respectively.

Exercise 6.17 Verify that (6.37) holds for hexahedral subparametric approximations in three dimensions, provided that the region R can be partitioned exactly.

6.3.2 Curved elements

Ciarlet and Raviart (1972b) have shown that error bounds such as (6.37) are valid for certain isoparametric elements with a *single curved side*. For curved elements, it is necessary to use the original form of (6.36) with $\|.\|_{k+1,R}$ on the right-hand side. It remains to verify Hypotheses 6.1 and 6.2.

(1) *Quadratic triangular elements* with one curved side can be written (Exercise 6.6) as

$$t = t_3 + (t_1 - t_3)p + (t_2 - t_3)q + \left(t_4 - \frac{t_1 + t_2}{2}\right)4pq \quad (t = x, y).$$

Since the interpolation is exact for quadratic polynomials in p and q, the bound on the error is

$$\| u - U \|_{1,R} = O(h^2)$$

provided that

$$\left\{ \left[x_4 - \frac{x_1 + x_2}{2} \right]^2 + \left[y_4 - \frac{y_1 + y_2}{2} \right]^2 \right\}^{1/2} = O(h^2). \tag{6.42}$$

This is equivalent to

$$\| P_4 - P_4^* \|_{\mathbb{R}^2} = O(h^2)$$

where P_4^* is the mid-point of the *chord* $P_1 P_2$ and the norm is the Euclidean distance in $\mathbb{R}^2$. This additional condition can always be satisfied if h is sufficiently small compared with the radius of curvature of the boundary. It is also assumed that the boundary can be represented *exactly* in terms of arcs that can be parametrized in the form

$$t = t_1 p(2p - 1) + t_2(1 - p)(1 - 2p) + t_4 4p(1 - p) \quad (t = x, y; p \in [0, 1]). \tag{6.43}$$

A similar bound holds for biquadratic isoparametric approximations based on quadrilaterals with a single curved side, if similar conditions are satisfied.

(2) *Cubic elements* can be written (Exercise 6.7) as

$$t = t_3 + (t_1 - t_3)p + (t_2 - t_3)q$$

$$+ 27pq(1 - p - q)\left(t_{10} - \frac{t_1 + t_2 + t_3}{3} \right)$$

$$+ \tfrac{9}{2}pq\left[(3p - 1)\left(t_4 - \frac{2t_1 + t_2}{3} \right) \right.$$

$$\left. + (3q - 1)\left(t_5 - \frac{2t_2 + t_1}{3} \right) \right] \quad (t = x, y).$$

Since the interpolation is exact for cubic polynomials in p and q, the bound on the error is then

$$\| u - U \|_{1,R} = O(h^3)$$

provided that

$$\| P_j - P_j^* \|_{\mathbb{R}^2} = O(h^2) \quad (j = 4, 5)$$

and

$$\| (P_4 - P_4^*) - (P_5 - P_5^*) \|_{\mathbb{R}^2} = O(h^3) \tag{6.44}$$

and that the point P_{10} is selected such that

$$t_{10} = \frac{t_1 + t_2 + t_3}{3} + \frac{(t_4 - t_4^*) + (t_5 - t_5^*)}{4} \quad (t = x, y) \tag{6.45}$$

where $P_4^* = (x_4^*, y_4^*)$ and $P_5^* = (x_5^*, y_5^*)$ are the points of trisection of the *chord* $P_1 P_2$ adjacent to P_4 and P_5 respectively. The additional inequality (6.44) is in fact a realistic condition if h is sufficiently small. As for the quadratic case, the bound on the finite element error is only valid if there is *no boundary perturbation*. A similar bound holds for Hermitian isoparametric approximation (Ciarlet and Raviart, 1972b), subject to a set of conditions analogous to (6.44) and (6.45).

These severe conditions on the curvature of isoparametric elements may be necessary in practice, as certain numerical evidence would indicate (Bond *et al.*, 1973)—but opinions differ.

In all the preceding convergence estimates in two dimensions it is assumed that θ, the smallest angle subtended by a triangular mesh at any node, is bounded away from zero as h tends to zero. In the analogous results for three dimensions, or for quadrilateral meshes in two dimensions, it is assumed that the ratio h/ρ remains bounded as h tends to zero. For any element, ρ is the diameter of the largest sphere (in $\mathbb{R}^3$)—circle in $\mathbb{R}^2$—that is contained in the element. If it is not possible to make such assumptions then the interpolation error bounds take the form

$$\|u - \tilde{u}\|_{r,R} = O\left(\frac{h^{k+1}}{\rho^r}\right) \tag{6.46}$$

assuming that $\|I - \Pi\|$ is uniformly bounded for all elements. This modified bound is introduced because in Hypothesis 6.1, the second inequality (6.46) is now in terms of ρ rather than h (Ciarlet and Raviart, 1972a; Bramble and Zlámal, 1970). For triangular meshes

$$\rho \approx h \sin \theta$$

hence the bound can be written in terms of $h^{k+1-r}/(\sin \theta)^r$. If it is not assumed that h/ρ is bounded as h tends to zero, then examples exist[†] for which $\|I - \Pi\|$ is not uniformly bounded and hence, as is shown by Bramble and Zlámal, the interpolation error bound can take the form

$$\|u - \tilde{u}\|_{r,R} \leqslant C\frac{h^{k+1-r}}{(\sin \theta)^{n+r}} |u|_{k+1,R}$$

for some $n \geqslant 1$, rather than (6.37). Babuška and Aziz (1976) have indicated that it is better to consider angles that tend to 2π rather than those that tend to zero, using the inequality $\rho \geqslant C \cos(\phi/2)$ where ϕ is the largest angle in the element.

[†] The 21-parameter C^1-quintic approximation is one such example.

Interpolation error bounds of the form

$$\|u - \tilde{u}\|_{r,R} \leqslant C_{k,r} h^{k+1-r} \tag{6.47}$$

have been derived using the *Sard kernel theorem* from methods based on both rectangles and triangles. For some forms of piecewise polynomial approximations, numerical values have been computed for the constants $C_{k,r}$ in (6.47). These results also suggest that bounds of the form (6.46) are not the best possible (Barnhill *et al.*, 1972, and references therein).

6.3.3 $\mathscr{L}_\infty$ Error bounds

The natural norm for error bounds in finite element calculations is the energy norm. In this norm the optimal properties of the Galerkin approximation are derived by a direct application of Cea's lemma. The $\mathscr{L}_2$ error bounds necessitated some additional manipulation via the Aubin–Nitsche lemma. The derivation of the $\mathscr{L}_\infty$ error bounds is much more technical and the detailed proofs involving *weighted Sobolev norms* will not be given. Interested readers should consult Nitsche (1979) for a good summary and an extensive bibliography.

A direct application of the obvious bound

$$\|u\|_{0,R} \leqslant \|u\|_{1,R}$$

leads to a non-optimal bound for the $\mathscr{L}_2$ error and, similarly, a direct application of the Sobolev lemma, i.e.

$$\|u\|_{(0)R} \leqslant C \begin{cases} \|u\|_{2,R} & \text{in } \mathbb{R}^2 \text{ or } \mathbb{R}^3. \\ \|u\|_{1,R} & \text{in } \mathbb{R}^1 \end{cases}$$

leads to non-optimal results. A number of different weighting functions $w(\mathbf{x})$ have been used to provide the best possible estimates. If $\mathbf{x}_0$ is a fixed reference point and

$$r = |\mathbf{x} - \mathbf{x}_0|$$

then the weight function can be written as

$$w(\mathbf{x}) = (\rho^2 + r^2)^{-\alpha} \quad \alpha, \rho > 0$$

(Nitsche, 1975, 1976; Natterer, 1977; Ciarlet, 1978).

Alternative weight functions have been used by Natterer (1975) and Babuška and Rosenweig (1972). The parameters α and ρ are chosen to provide the desired order of accuracy. The proof establishes the equivalence of the weighted Sobolev norm,

$$\|u\|_{k,R;w} = \left\{ \sum_{|j| \leqslant k} \int_R (D^j u)^2 w \, d\mathbf{x} \right\}^{1/2}$$

to the standard Sobolev norm $\|u\|_{k,R}$. It is then possible to use the results in the preceding sections with an additional step using the Sobolev lemma converting the bounds into an estimate of the form

$$|w(x_0)(u - U)(x_0)| = O(h^k).$$

The values of the parameters α and ρ now become important. If $\rho = O(h)$ then the bound becomes

$$|(u - U)(x_0)| = O(h^{k + 2\alpha}).$$

It is found that the proof is only valid for

$$0 < \alpha < \tfrac{1}{2}$$

for linear elements and

$$0 < \alpha \leqslant \tfrac{1}{2}$$

for higher orders.

Lemma 6.8 For second-order problems

$$\|u - U\|_{(0)R} \leqslant Ch^{2-\varepsilon}|u|_{(2)R} \tag{6.48}$$

for linear elements and

$$\|u - U\|_{(0)R} \leqslant Ch^{k+1}|u|_{(k+1)R} \tag{6.49}$$

for piecewise polynomials of degree k.

The proof can be found in Natterer (1975) or Nitsche (1975). There is an alternative representation of the bound in the case of piecewise linears.

Corollary (Nitsche, 1976) *For piecewise linear elements*

$$\|u - U\|_{(0)R} \leqslant h^2|\ln h|^{1/2}|u|_{(2)R} \tag{6.50}$$

(see Ciarlet (1978) for a proof of a weaker result with $|\ln h|^{3/2}$).

The existence of the factor $|\ln h|$ is linked to the form of the Green's function (see Stakgold, 1979) which, for Laplace's equation in $\mathbb{R}^2$, is

$$G(\mathbf{x}, \mathbf{x}_0) = \ln|\mathbf{x} - \mathbf{x}_0|.$$

Scott (1976) has obtained the same results as quoted above by an alternative analysis that considers the Green's function explicitly. Examples provided by Fried (1980) and Jespersen (1978) have shown that the bound (6.50) is the best possible and is not just a consequence of the method adopted in the proof.

In a series of papers Schatz and Wahlbin (1977, 1978, 1979, 1981, 1982) show that the $\mathscr{L}_\infty$ norm is seriously affected by the lack of smoothness of the solution in the neighbourhood of re-entrant corners. Any singularity in the boundary data leads to a marked deterioration in the accuracy locally and this is highlighted in the $\mathscr{L}_\infty$ bound. The so-called *interior estimates* are an attempt to measure the accuracy away from the polluting effect of the corners and to recapture the exponent in (6.48).

6.4 Perturbation Errors

The error bounds (6.32) and (6.33) involve two distinct types of error:

(1) The first term

$$\inf_{V_h \in K_h} \|u - V_h\|$$

is an *approximation error* and can be bounded by the methods outlined in Section 6.3.1.
(2) All the remaining terms are introduced because of the perturbed form of the Galerkin equation (6.28). In this section we seek bounds on these perturbation terms for different forms of the perturbation.

The finite element solution is said to be *optimal* if the order of the perturbation errors are no bigger than that of the approximation error (Nitsche, 1972).

In their analysis of quadrature perturbations, Herbold and Varga (1972) call the quadrature scheme *consistent* if the resulting approximation is optimal.

6.4.1 Numerical integration

Bounds on the errors in integrals evaluated numerically by means of a standard quadrature rule are considered in Exercises 6.8 and 6.9 in Section 6.1. In this section we consider such quadratures in more detail and provide bounds that are valid for certain types of nonlinear transformation from T_0 onto T.

A quadrature scheme on the standard element is defined as a set of points $\mathbf{p}_l \in T_0 (l = 1, \ldots, L)$ and a set of positive weights $b_l (l = 1, \ldots, L)$. The condition $b_l > 0$ is necessary if the perturbed bilinear form a_h is to be K_h-elliptic. Any integral on the standard element can be written as

$$\iint_{T_0} u(\mathbf{p})\,\mathrm{d}\mathbf{p} = \sum_{l=1}^{L} b_l u(\mathbf{p}_l) + E_0(u)$$

where—as before—E_0 is the quadrature error operator for the standard element. To transform the quadrature scheme onto an arbitrary element we take the points $\mathbf{x}_l = \mathbf{F}(\mathbf{p}_l) \in T$ and the weights $\beta_l = b_l J(\mathbf{p}_l)$, where J is the Jacobian of the transformation $\mathbf{F}$. If we denote the quadrature error in T by $E(u)$, it follows that

$$E(u) = \iint_T u(\mathbf{x})\,d\mathbf{x} - \sum_{l=1}^{L} \beta_l u(\mathbf{x}_l)$$

$$= \iint_{T_0} u(\mathbf{p})J(\mathbf{p})\,d\mathbf{p} - \sum_{l=1}^{L} b_l J(\mathbf{p}_l)u(\mathbf{p}_l)$$

$$= E_0(uJ).$$

In this section we follow the analysis of Ciarlet and Raviart (1972c) and study the perturbation errors in the solution of the differential equation

$$\frac{\partial}{\partial x}\left(d_1 \frac{\partial u}{\partial x}\right) + \frac{\partial}{\partial y}\left(d_2 \frac{\partial u}{\partial y}\right) + f(x, y) = 0 \quad ((x, y) \in R) \tag{6.51}$$

subject to

$$u(x, y) = 0 \quad ((x, y) \in \partial R).$$

It is assumed that the perturbations are due entirely to evaluating the inner products by numerical quadrature. Thus all the basis functions φ satisfy the boundary condition and are conforming—in this problem $\varphi \in \mathcal{H}^{(1)}(R)$. It is further assumed that the region R is partitioned into S elements $T_j(j = 1, \ldots, S)$, and that corresponding to each element there is a transformation $\mathbf{F}_j$ from the standard element; the Jacobian of $\mathbf{F}_j$ is denoted by J_j.

The bilinear form corresponding to (6.51) is therefore

$$a(u, v) = \sum_{j=1}^{S} \iint_{T_j}\left\{d_1\left(\frac{\partial u}{\partial x}\right)\left(\frac{\partial v}{\partial x}\right) + d_2\left(\frac{\partial u}{\partial y}\right)\left(\frac{\partial v}{\partial y}\right)\right\}dxdy$$

and the perturbed form is

$$a_h(u, v) = \sum_{j=1}^{S} \sum_{l=1}^{L} \beta_{ij}\left\{d_1\left(\frac{\partial u}{\partial x}\right)\left(\frac{\partial v}{\partial x}\right) + d_2\left(\frac{\partial u}{\partial y}\right)\left(\frac{\partial v}{\partial y}\right)\right\}_{\mathbf{x}=\mathbf{F}_j(\mathbf{p}_l)}$$

where

$$\beta_{ij} = b_l J_j(\mathbf{p}_l) \quad \begin{cases} j = 1, \ldots, S \\ l = 1, \ldots, L. \end{cases}$$

Thus

$$(a - a_h)(u, v) = \sum_{j=1}^{S}\left\{E_0\left(d_1 \frac{\partial u}{\partial x}\frac{\partial v}{\partial x}J_j\right) + E_0\left(d_2 \frac{\partial u}{\partial y}\frac{\partial v}{\partial y}J_j\right)\right\}. \tag{6.52}$$

The individual terms in the summation are thus of the form $E_0(zw)$, where $z(\mathbf{p})$ and $w(\mathbf{p})$ are $d_1(\partial u/\partial x)$ and $(\partial v/\partial x)J_j$ respectively (or $d_2(\partial u/\partial y)$ and $(\partial v/\partial y)J_j$) and we assume that d_1 and d_2 are sufficiently differentiable to ensure that $z \in \mathcal{H}^{(k)}(T_0)$.

Then it is possible to use the bound on integrals of products derived in Exercise 6.9 provided $(\partial v/\partial x)J_j$ and $(\partial v/\partial y)J_j$ are polynomials in $\mathbf{p}$.

Similarly the quadrature scheme is applied to the right-hand side of the Galerkin equation

$$a(u, v) = (f, v)$$

where

$$(f, v) = \sum_{j=1}^{S} \int\!\!\int_{T_j} f(\mathbf{x})v(\mathbf{x})\,d\mathbf{x}.$$

Thus it follows that

$$(f, v)_h = \sum_{j=1}^{S} \sum_{l=1}^{L} \beta_{lj}\{f(\mathbf{x})v(\mathbf{x})\}_{\mathbf{x} = F_j(\mathbf{p}_l)}$$

and hence

$$(f, v) - (f, v)_h = \sum_{j=1}^{S} E_0(fvJ_j). \tag{6.53}$$

Again if $f \in \mathscr{H}^{(k)}(T_0)$ for some k, it is possible to apply the quadrature error bounds to this form of product provided that vJ_j is a polynomial in $\mathbf{p}$. Once we have obtained bounds on the perturbations (6.52) and (6.53) these bounds can be used in (6.31) or (6.33) to estimate the convergence of the approximation.

If for each element the transformation from the standard element is linear, then the Jacobians J_j are all constant and if $W_h(\mathbf{x})$ is a polynomial so are $\partial W_h(\mathbf{p})/\partial x$, $\partial W_h(\mathbf{p})/\partial y$, and $W_h(\mathbf{p})$. It can be shown that in each element the functions $\partial W_h(\mathbf{p})/\partial x J(\mathbf{p})$, $\partial W_h(\mathbf{p})/\partial y J(\mathbf{p})$ and $W_h(\mathbf{p})J(\mathbf{p})$ are all polynomials when $W_h(\mathbf{x})$ is a finite element trial function of various kinds (Ciarlet and Raviart, 1972c).

Given the assumption on the polynomial nature of $J(\mathbf{p})\partial W_h/\partial x$, $J(\mathbf{p})\partial W_h/\partial y$, and $J(\mathbf{p})W_h$ in each element, it is possible to derive bounds on the perturbations (6.52) and (6.53).

Theorem 6.6 Assume that for any trial function $W_h \in K_h$ it follows that in each element, $J(\mathbf{p})\partial W_h/\partial x$ and $J(\mathbf{p})\partial W_h/\partial y$ are polynomials of degree at most r_1, that $J(\mathbf{p})W_h$ is a polynomial of degree at most r_0, and that the regularity hypothesis is valid for all $s \leqslant \max\{r_1, r_0\}$. Then if the quadrature scheme (on the standard triangle) is exact for all polynomials of degree at most $r_1 + s$, it follows that the perturbation (6.52) in the bilinear form is bounded as

$$\frac{|(a - a_h)(V_h, W_h)|}{\|W_h\|_{1,R}} \leqslant Ch^{s+1}\|V_h\|_{s+2,R} \tag{6.54}$$

where V_h, $W_h \in K_h$ are any trial functions. Similarly, if the quadrature is exact for all *polynomials of degree at most* $r_0 + s - 1$, *then the perturbation* (6.53) *in the right-hand side is bounded as*

$$\frac{|(f, W_h) - (f, W_h)_h|}{\| W_h \|_{1,R}} \leqslant Ch^{s+1} \| f \|_{s+1,R} \tag{6.55}$$

for any $W_h \in K_h$ *provided that* $f \in \mathcal{H}^{(s+1)}(R)$.

If, for example, we use *Lagrange or Hermite elements of degree* k, the transformation from T to T_0 is linear and in each element $J_0(\partial W_h / \partial x) \in P_{k-1}$, $J_0(\partial W_h / \partial y) \in P_{k-1}$, and $J_0 W_h \in P_k$. Thus $r_1 = k - 1$, $r_0 = k$ and hence if we use a quadrature that is *exact for all polynomials of degree at most* $2k - 2$, it follows that $s = k - 1$ in (6.54) and (6.55). It then follows from (6.33) that

$$\| u - U_h \|_{1,R} \leqslant C \| u - \tilde{u} \|_{1,R} + \sup_{W_h \in K_h} \left\{ \frac{|(a - a_h)(\tilde{u}, W_h)|}{\| W_h \|_{1,R}} \right\}$$

$$+ \sup_{W_h \in K_h} \left\{ \frac{|(f, W_h) - (f, W_h)_h|}{\| W_h \|_{1,R}} \right\}$$

where $\tilde{u}$ interpolates u.

Corollary Following the arguments in Section 6.3, Theorem 6.6 implies that

$$\| u - \tilde{u} \|_{1,R} \leqslant Ch^k |u|_{k+1,R}$$

and the Galerkin approximation is optimal, i.e.

$$\| u - U_h \|_{1,R} = O(h^k)$$

since the perturbation terms (6.54) *and* (6.55) *are also* $O(h^k)$.

The theorem not only shows the degree of quadrature rule that is necessary to give an optimal approximation; it also shows that the minimum degree necessary to ensure convergence as h tends to zero is min $\{r_1, r_0 - 1\}$. In the above example this would be $k - 1$.

Proof of Theorem 6.6 The error in the bilinear form is made up of terms such as

$$E_0 \left(d_1 \frac{\partial V_h}{\partial x} \frac{\partial W_h}{\partial x} J \right) = E_0(vw)$$

where $v(\mathbf{p}) = d_1(\partial V_h / \partial x)$ and $w(\mathbf{p}) = J(\partial W_h / \partial x)$. If $w \in P_{r_1}$ and the quadrature is exact for polynomials of degree $r_1 + s$, it follows from the Bramble–Hilbert lemma that

$$|E_0(vw)| \leqslant C \| w \|_{0,T_0} |v|_{s+1,T_0}.$$

Assuming that d_1 is sufficiently differentiable, it follows from (6.9a) and

Exercise 6.4 that

$$|E_0(vw)| \leqslant Ch^{s+1} \left\| \frac{\partial W_h}{\partial x} \right\|_{0,T} \left\| \frac{\partial V_h}{\partial x} \right\|_{s+1,T}$$

$$\leqslant Ch^{s+1} \|W_h\|_{1,T} \|V_h\|_{s+2,T}.$$

By summing over all elements and dividing by $\|W_h\|_{1,R}$ we obtain (6.54).

In order to obtain (6.55) we follow the proof of Ciarlet and Raviart (1972c) and introduce a projection Π_0 onto the space P_0 of constant functions on T_0; thus for any $u \in \mathscr{L}_2(T_0)$ it follows that

$$\iint_{T_0} (u - \Pi_0 u) \, \mathrm{d}\mathbf{p} = 0.$$

A typical term in the error in the right-hand side of the Galerkin equations can be written as

$$E_0(fW_hJ) = E_0(fJ[I - \Pi_0]W_h) + E_0(fJ[\Pi_0 W_h]).$$

Since $J[I - \Pi_0]W_h \in P_{r_0}$, it follows from the Bramble–Hilbert lemma (Exercise 6.9) that

$$|E_0(fJ[I - \Pi_0]W_h)| \leqslant C \|J[I - \Pi_0]W_h\|_{0,T_0} |f|_{s,T_0}.$$

But Π_0 is a projection operator and so it is possible to apply Lemma 6.6 to obtain

$$\|[I - \Pi_0]W_h\|_{0,T_0} \leqslant C|W_h|_{1,T_0}.$$

It follows from (6.9a), therefore, that

$$|E_0(fJ[I - \Pi_0]W_h)| \leqslant C \left\{ \sup_{\mathbf{p} \in T_0} J(\mathbf{p}) \right\} |f|_{s,T_0} |W_h|_{1,T_0}$$

$$\leqslant Ch^{s+1} \|f\|_{s,T} \|W_h\|_{1,T}. \tag{6.56}$$

Similarly as $J(\mathbf{p}) \in P_{r_0}$ and $\Pi_0 W_h$ is a constant it follows that

$$|E_0(fJ[\Pi_0 W_h])| \leqslant C \|J[\Pi_0 W_h]\|_{0,T_0} |f|_{s+1,T_0}$$

$$\leqslant C \left\{ \sup_{\mathbf{p} \in T_0} J(\mathbf{p}) \right\} \|W_h\|_{0,T_0} |f|_{s+1,T_0}$$

$$\leqslant Ch^{s+1} \|f\|_{s+1,T} \|W_h\|_{0,T}. \tag{6.57}$$

Combining (6.56) with (6.57), summing over all elements, and dividing by $\|W_h\|_{1,R}$ leads to the desired result.

Results similar to Theorem 6.6 have been obtained by Fix (1972) in the study of the effect of quadrature formulae with both Lagrangian and Hermitian finite element approximations for a polygonal region. Quadrature schemes have also been studied by Herbold and Varga (1972), but only for rectangular regions and assuming that the bilinear form a_h was integrated exactly.

Ciarlet and Raviart (1972c) have applied the results of Theorem 6.6 to isoparametric approximations defined in terms of both triangles and quadrilaterals. They also show how quadrature schemes can be chosen such that the bilinear form a_h is K_h-elliptic and hence it is possible to justify the use of the error estimate (6.33).

These results, however, lead to useful error bounds only when isoparametric elements are used with, at most, *one curved side*. Even in such cases the results are subject to the conditions outlined in Section 6.3. Most of the quadrature error bounds generalize to $\mathbb{R}^m (m > 2)$ for differential equations of the form

$$\sum_{i,j=1}^{M} \frac{\partial}{\partial x_i}\left(d_{ij} \frac{\partial u}{\partial x_j} \right) + f(\mathbf{x}) = 0$$

where $\mathbf{x} = (x_1, \ldots, x_m)^{\mathrm{T}}$.

Exercise 6.18 Verify that, for the quadratic isoparametric element given in Exercise 6.6, it follows that $J(\partial p/\partial x)$, $J(\partial p/\partial y)$, $J(\partial q/\partial x)$ and $J(\partial q/\partial y)$ are all *linear* functions of p and q. Hence verify that for any piecewise quadratic trial function W_h, it follows that in each triangle the functions $J(\mathbf{p})\partial W_h(\mathbf{p})/\partial x$ and $J(\mathbf{p})\partial W_h(\mathbf{p})/\partial y$ are *quadratic* polynomials in $\mathbf{p}$; also verify that $J(\mathbf{p})W_h(\mathbf{p})$ is a fourth-order polynomial.

Exercise 6.19 Prove that if triangular isoparametric elements of degree k are used in each triangle; $J, J(\partial W_h/\partial x)$ and $J(\partial W_h/\partial y)$ are all polynomials of degree $2(k-1)$ for any trial function W_h. Prove that, in general, $JW_h \in P_{3k-2}$ in each triangle.

Exercise 6.20 By considering errors of the form

$$E_0\left(\frac{\partial^2 V_h}{\partial x^2} \frac{\partial^2 W_h}{\partial x^2} J \right)$$

show that if the quadrature is exact for polynomials of degree $r_1 + s$ and the transformation from an arbitrary element onto the standard element is *linear* then the perturbation $(a - a_h)$ for fourth-order problems can be bounded as

$$\frac{(a - a_h)(V_h, W_h)}{\| W_h \|_{2,R}} \leqslant Ch^{s+1} \| V_h \|_{s+3,R}$$

provided the trial functions are conforming and are polynomials of degree at most $r_1 + 2$ in each element.

Exercise 6.21 Using the results of Exercise 6.19, verify that an isoparametric approximation of degree k is optimal if a quadrature rule of degree $4(k-1)$ is used.

6.4.2 Interpolated boundary conditions

We now assume that the approximating subspace K_h contains functions that do not vanish on the boundary but that the integrals are evaluated exactly. Thus it is possible to use the error bound (6.33) in which the perturbation term is

$$\sup_{W_h \in K_h} \left\{ \frac{|(f, W_h) - a(u, W_h)|}{\| W_h \|} \right\}. \tag{6.58}$$

We consider initially the approximate solution of

$$a(u, v) = (f, v)$$

where

$$a(u, v) = \int\int_R \left\{ \left(\frac{\partial u}{\partial x} \right) \left(\frac{\partial v}{\partial x} \right) + \left(\frac{\partial u}{\partial y} \right) \left(\frac{\partial v}{\partial y} \right) \right\} dx dy$$

corresponding to the differential equation

$$\frac{\partial^2 u}{\partial x^2} + \frac{\partial^2 u}{\partial y^2} + f(x, y) = 0 \quad ((x, y) \in R)$$

subject to

$$u = 0 \quad ((x, y) \in \partial R).$$

The results of this section can be applied with inhomogeneous boundary conditions if, as in Section 6.1, a suitable continuation of the boundary data is available. An alternative approach might be to use Theorem 6.3 to estimate the errors.

An analysis of methods that do not satisfy the boundary conditions exactly nearly always involves boundary integrals. This is also true of the penalty methods described in Section 6.4.4. It is possible to define a Sobolev space $\mathscr{H}^{(k-1)}(\partial R)$ analogous to the space $\mathscr{H}^{(k-1)}(R)$ defined in Section 6.1. It follows from such a definition that, for example,

$$\left| \int_{\partial R} \left(\frac{\partial u}{\partial n} \right) W_h \, ds \right| \leqslant \left\| \frac{\partial u}{\partial n} \right\|_{k-1, \partial R} \| W_h \|_{1-k, \partial R}.$$

Then, by the trace theorem, it follows that

$$\left\| \frac{\partial u}{\partial n} \right\|_{k-1, \partial R} \leqslant C \| u \|_{k+1, R}.$$

It then follows from Green's theorem that

$$|(f, W_h) - a(u, W_h)| = \left| \int_{\partial R} \left(\frac{\partial u}{\partial n} \right) W_h \, ds \right|$$

and hence, by combining these two expressions, we obtain bounds of the form

$$|(f, W_h) - a(u, W_h)| \leqslant C \| u \|_{k+1, R} \| W_h \|_{1-k, \partial R}.$$

To be of any use in second-order problems, (6.58) must involve $\| W_h \|_{1, R}$ and not

$\| W_h \|_{1-k,\partial R}$. For this reason Scott (1975) has derived bounds of the form

$$\sup_{W_h \in K_h} \left\{ \frac{\| W_h \|_{1-k,\partial R}}{\| W_h \|_{1,R}} \right\} \leqslant Ch^{k+1/2} \tag{6.59}$$

for trial functions W_h that are *nearly zero* on the boundary ∂R. Berger, *et al.* (1972) proved (6.59) for the particular case $k = 1$, but as they did not make the best possible choice of trial functions they were unable to generalize the result. From the definitions of a dual space, it follows that (6.59) is equivalent to

$$\sup_{W_h \in K_h} \left\{ \sup_{g \in \mathscr{H}^{(k-1)}(\partial R)} \left\{ \frac{|\int_{\partial R} g W_h \, ds|}{\| g \|_{k-1,\partial R} \| W_h \|_{1,R}} \right\} \right\} \leqslant Ch^{k+1/2}$$

that is, for all $g \in \mathscr{H}^{(k-1)}(\partial R)$ and $W_h \in K_h$

$$\left| \int_{\partial R} g W_h \, ds \right| \leqslant Ch^{k+1/2} \| g \|_{k-1,\partial R} \| W_h \|_{1,R} \tag{6.60}$$

and hence

$$\left\{ \frac{|(f, W_h) - a(u, W_h)|}{\| W_h \|_{1,R}} \right\} \leqslant Ch^{k+1/2} \| u \|_{k+1,R}. \tag{6.61}$$

Theorem 6.5 shows that Lagrange interpolating polynomials of degree k lead to interpolation error bounds of the form

$$\| u - \tilde{u} \|_{1,R} \leqslant Ch^k |u|_{k+1,R}$$

and hence the Galerkin error bounds are of the form

$$\| u - U_h \|_{1,R} \leqslant C\{h^k |u|_{k+1,R} + h^{k+1/2} \| u \|_{k+1,R}\}. \tag{6.62}$$

So a finite element approximation with such interpolated boundary conditions is *optimal* since the perturbation error is of a higher order (in h) than the approximation error. Scott (1975) and Chernuka *et al.* (1972) have devised quadrature rules for triangular elements with curved boundaries that preserve the order for piecewise quadratic approximations. Piecewise quadratics have also been studied by Berger (1973) to derive a bound on the error in terms of the $\mathscr{L}_2(R)$ norm; he has also verified the orders numerically (1972). As an alternative to interpolating the boundary data at a finite point set, the approximation could be made to match along the entire boundary if blending function interpolants are used (Gordon and Wixom, 1974).

6.4.3 Boundary approximation

Probably the first error bounds for finite element methods over perturbed regions were obtained by Russian mathematicians (for example Oganesyan, 1966). They derived bounds for piecewise linear approximations based on triangular meshes and they considered only the approximate solution of second-order problems subject to the boundary condition

$$\frac{\partial u}{\partial n} + \beta u = 0 \quad (\beta \geqslant 0)$$

on a curved boundary. They showed that

$$\| u - U_h \|_{1,R_h} \leqslant Ch \| u \|_{2,R_h}$$

but the proofs are rather technical and are beyond the scope of this book. Others (Oganesyan and Rukhovets, 1969) have also derived error bounds for this type of problem in terms of the $\mathscr{L}_2(R)$ norm. More recently, it has been shown (Strang and Berger, 1971; Thomée, 1973) that if R_h is a polygon inscribed in $R \subset \mathbb{R}^2(\mathbb{R}^m, m \geqslant 2$ according to Strang and Fix, 1973, p. 196) then for the model problem of

$$\frac{\partial^2 u}{\partial x^2} + \frac{\partial^2 u}{\partial y^2} + f(x, y) = 0 \quad ((x, y) \in R) \tag{6.63}$$

subject to $u = 0$ on ∂R, it follows that

$$\| u - u_h \|_{1,R_h} = O(h^{3/2})$$

where u_h is the solution of the perturbed problem consisting of

$$\frac{\partial^2 u_h}{\partial x^2} + \frac{\partial^2 u_h}{\partial y^2} + f(x, y) = 0 \quad ((x, y) \in R_h) \tag{6.64}$$

subject to $u_h = 0$ on ∂R_h.

Thus if (6.63) is solved approximately, by partitioning the polygonal region R_h and then computing a finite element solution of (6.64), it follows from Section 6.3 that

$$\| u - U_h \|_{1,R_h} = O(h)$$

for piecewise linear approximations based on a triangular partition. Similarly it follows that if approximating functions include all polynomials of degree 2, or higher, then

$$\| u - U_h \|_{1,R_h} = O(h^{3/2}).$$

This order of approximation may be significantly lower than that expected from Section 6.3 and it arises from a poor approximation near the boundary; sometimes referred to as a *boundary layer effect*. Maximum principles can be used to verify that the perturbations are smaller in the interior of R, where sharper bounds are available. Some of these results can be extended to the situation when $R_h \not\subset R$. The convergence properties of the finite element approximations away from the boundary have also been studied by Nitsche and Schatz (1974) and Bramble and Thomée (1974).

Berger *et al.* (1972) show that, in general, if the region R is approximated by R_h—which is not necessarily a polygon—such that the maximum distance between the two boundaries ∂R and ∂R_h is $O(h^{k+1})$ then the perturbation term in (6.32) is $O(h^{k+1/2})$. As a piecewise k-degree polynomial approximation of the boundary—by interpolation say—could satisfy this condition, it follows that if the same degree of polynomials are used in both the function approximation and the boundary interpolation, then the finite element error is still $O(h^k)$ overall in terms of the $\mathscr{H}_2^{(1)}(R_h)$ norm. An analogous result is true for isoparametric approximations since Ciarlet and Raviart (1972c) have shown that the conclusions of Theorem 6.6, subject to minor modifications, are valid when the region is perturbed. A similar result has been obtained by Zlámal (1973, 1974). The Neumann problem has been considered by a few authors such as Strang and Fix (1973) and Babuška (1971).

6.4.4 Penalty methods

This category includes all methods that incorporate nonhomogeneous Dirichlet boundary conditions in the form of a boundary integral that is added to the functional, rather than as a condition to be imposed on the approximating functions. Such methods can be based on the method of least squares or the Ritz method, or a combination of both. The most frequently used formulation is based on the method of least squares for which the errors are no longer derived naturally in terms of Sobolev norms and invariably involve the *trace theorem* to deal with the boundary integrals (for example Varga, 1971, Chapter 6).

If we assume that the interpolation error bound given by Theorem 6.5 is valid for some $k > 0$, that is,

$$\| u - \tilde{u} \|_{r,R} \leqslant Ch^{k+1-r} \| u \|_{k+1,R}$$

then it is possible to derive a straightforward energy bound on the error—but not in terms of Sobolev norms.

Theorem 6.7 *If the finite element approximation satisfies*

$$(AU_h - f, AV_h) = h^{-3} \langle U_h - g, V_h \rangle \quad (\textit{for all } V_h \in K_h) \tag{6.65}$$

where

$$\langle \varphi, \psi \rangle = \int_{\partial R} \varphi \psi \, \mathrm{d}s$$

then if Theorem 6.5 is valid for some $k > 0$, it follows that

$$\| AU_h - Au \|_{0,R} + h^{3/2} \| U_h - u \|_{0,\partial R} \leqslant Ch^{k-1} \| u \|_{k+1,R}.$$

Proof Since

$$\| Au - A\tilde{u} \|_{0,R} \leqslant C \| u - \tilde{u} \|_{2,R}$$

for any $(u - \tilde{u}) \in \mathcal{H}^{(2)}(R)$ and

$$\|u - \tilde{u}\|_{0,\partial R} \leqslant C\{h^{-1/2}\|u - \tilde{u}\|_{0,R} + h^{1/2}\|u - \tilde{u}\|_{1,R}\}$$

Agmon (1965), it follows that

$$\|Au - A\tilde{u}\|_{0,R} + h^{3/2}\|u - \tilde{u}\|_{0,R}$$

$$\leqslant C\{\|u - \tilde{u}\|_{2,R} + h^{-1}\|u - \tilde{u}\|_{1,R} + h^{-2}\|u - \tilde{u}\|_{0,R}\}. \tag{6.66}$$

Since the least squares method given by (6.65) is a *projection method* in the sense that it is possible to apply Cea's lemma as

$$\|u - U_h\|_{[1]} = \inf_{\tilde{u} \in K_h} \|u - \tilde{u}\|_{[1]}$$

where the norm is defined by

$$\|\varphi\|_{[1]}^2 = \|A\varphi\|_{0,R}^2 + h^{-3}\|\varphi\|_{0,\partial R}^2.$$

The result follows immediately from (6.66) and Theorem 6.5 with $r = 0, 1$ and 2.

It is also possible to prove that

$$\|u - U_h\|_{0,R} \leqslant Ch^{k+1}\|u\|_{k+1,R}$$

but the proof is beyond the scope of this book (Baker, 1973; Bramble and Schatz, 1970).

Other authors have suggested alternative projection methods for solving (6.63); they have based their methods on such norms as

$$\|\varphi\|_{[2]}^2 = a(\varphi, \varphi) + h^{-1}\|\varphi\|_{0,\partial R}^2$$

(Bramble *et al.* 1972) and

$$\|\varphi\|_{[3]}^2 = -a(\varphi, \varphi) - 2(A\varphi, \varphi) + h^2\|A\varphi\|_{0,R}^2$$

$$+ \gamma\left\{h^{-1}\|\varphi\|_{0,\partial R}^2 + h\left\|\frac{\partial\varphi}{\partial s}\right\|_{0,\partial R}^2\right\} \quad (\gamma > 0)$$

(Bramble and Nitsche, 1973).

Such methods have been extended to problems of higher degree and also to problems in more than two dimensions. Methods based on stationary points of functionals that are not positive definite have also been suggested (for example Thomée, 1973). Penalty methods have also been studied by Aubin (1972, p. 17).

6.4.5 Nonconforming elements

Let K_h be the approximating space of nonconforming basis functions. Then a nonconforming Galerkin approximation $U_h \in K_h$ satisfies

$$a_h(U_h, V_h) = (f, V_h), \quad \text{for all } V_h \in K_h \tag{6.67}$$

where

$$a_h(u, v) = \sum_{j=1}^{S} \int\int_{T_j} \left\{ \frac{\partial u}{\partial x} \frac{\partial v}{\partial x} + \frac{\partial u}{\partial y} \frac{\partial v}{\partial y} \right\} dx\, dy$$

does not equal

$$a(u, v) = \int\int_{R} \left\{ \frac{\partial u}{\partial x} \frac{\partial v}{\partial x} + \frac{\partial u}{\partial y} \frac{\partial v}{\partial y} \right\} dx\, dy$$

if v has a jump discontinuity across the inter-element boundaries. It is the difference between the two forms a and a_h that is crucial in the study of nonconforming elements. In addition we define the energy semi-norm corresponding to a_h by

$$|u|_h = [a_h(u, u)]^{1/2}$$

and also a norm

$$\|u\|_h = [|u|_h^2 + \|u\|_{0,R}^2]^{1/2}.$$

As an example of the results that can be obtained for nonconforming elements we consider the space K_h of piecewise linear elements that are matched at the *side midpoints* of a triangular mesh. It follows with such elements that the error in interpolating linear functions is zero, so by analogy with Section 6.3:

$$\inf_{V_h \in K_h} \|u - V_h\|_h \leqslant Ch|u|_{2,R}$$

for any $u \in \mathscr{H}^{(2)}(R)$. Thus if the nonconforming approximation is to be optimal, the perturbation term in (6.32),

$$\sup_{W_h \in K_h} \left\{ \frac{|(f, W_h) - a_h(u, W_h)|}{\|W_h\|_h} \right\} \tag{6.68}$$

must be at least $O(h)$.

By applying Green's theorem in each element it follows that

$$(f, W_h) - a_h(u, W_h) = \sum_{j=1}^{S} \int_{\partial T_j} \frac{\partial u}{\partial n} W_h \, ds$$

$$= \sum_{l=1}^{Q} \int_{E_l} \left\{ \left(\frac{\partial u}{\partial n} W_h \right)^{[1]} + \left(\frac{\partial u}{\partial n} W_h \right)^{[2]} \right\} ds$$

where E_l denotes a particular edge of the mesh and Q is the number of edges in the mesh. The two terms in the final integral, numbered [1] and [2] respectively, are the limit values corresponding to the elements on either side of the interface E_l. By applying the Bramble–Hilbert lemma to functionals of the form

$$F(u, W_h) = \int_{E_l} \left\{ \left(\frac{\partial u}{\partial n} W_h \right)^{[1]} + \left(\frac{\partial u}{\partial n} W_h \right)^{[2]} \right\} ds. \tag{6.69}$$

(Exercise 6.23), it is possible to show (Crouzeix and Raviart, 1973) that if

$$\int_{E_l} \{W_h^{[1]} - W_h^{[2]}\}\, ds = 0 \tag{6.70}$$

for all $W_h \in K_h$, then

$$|(f, W_h) - a_h(u, W_h)| \leqslant Ch|u|_{1,R}\|W_h\|_h. \tag{6.71}$$

Ciarlet (1973a) has produced similar results for plate-bending elements. A comprehensive analysis of nonconforming methods for plate-bending problems has been provided by Lascaux and Lesaint (1975).

As an alternative to the preceding analysis, Irons (Irons and Razzaque, 1972) introduced a simple *patch test* as a means of testing the suitability of nonconforming elements. It can be stated as follows. Suppose the space of nonconforming basis functions contains all polynomials of the same degree r as the highest derivative in the energy norm and that round the perimeter of any arbitrary patch of elements the boundary conditions are chosen consistent with a particular solution $u = p_r \in P_r$ within the patch. The patch test then requires the approximation U_h, calculated by the Galerkin finite element method, *ignoring discontinuities at the element interfaces, to coincide with p_r within the patch*. The patch test thus requires that if $u = p_r \in P_r$, then U_h calculated from (6.67) must satisfy

$$U_h = p_r \tag{6.72}$$

and so in this particular case,

$$a(p_r, V_h) = a_h(p_r, V_h) \quad \text{for all } V_h \in K_M. \tag{6.73}$$

The justification for accepting that

$$a(u, V_h) = (f, V_h)$$

at $u = p_r \in P_r$ although $K_M \not\subset \mathscr{H}$, is given by Strang (1972). The statements implicit in (6.72) and (6.73) are alternative but equivalent forms of the patch test.

If the patch test is passed, we can say that the perturbation term (6.68) is zero for the test solution $u = p_r \in P_r$.

It had been thought that passing the patch test might be a necessary and/or sufficient condition for the convergence of a Galerkin approximation based on nonconforming elements. However, a counterexample, albeit one-dimensional and contrived, recently produced by Stummel (1980) has negated that hope. All we can say at the moment is that if a particular nonconforming element passes the patch test, it is likely to be of some practical value to the user. Alternatively, the patch test can be thought of as a test for *consistency* (Irons and Loikkanen, 1983). The much tougher problem of convergence for nonconforming elements, however, appears to be a separate issue.

As an illustration of the patch test for second-order problems, consider the

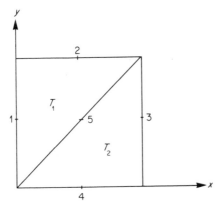

Figure 6.2

solution of Laplace's equation in a unit square patch comprising two triangular elements (Figure 6.2). The energy norm contains first-order derivatives and so $r = 1$. In each triangle we choose a linear function that matches the function values at the midpoints of the three sides. This leads to

$$U^{[1]}(x, y) = (1 - 2x)U_1 - (1 - 2y)U_2 + (1 + 2x - 2y)U_5$$

and

$$U^{[2]}(x, y) = -(1 - 2x)U_3 + (1 - 2y)U_4 + (1 - 2x + 2y)U_5$$

in triangles T_1 and T_2 respectively. The overall interpolant is not in general continuous across the interface of the two triangles and so this element is nonconforming for the Galerkin method.

Consider for example the test solution within the patch to be

$$u = x + y$$

leading to the boundary values

$$u_1 = u_4 = \tfrac{1}{2} \quad u_2 = u_3 = \tfrac{3}{2}$$

and to the modified interpolants

$$U^{[1]}(x, y) = (-1 - x + 3y) + (1 + 2x - 2y)U_5 \tag{6.74a}$$

and

$$U^{[2]}(x, y) = (-1 + 3x - y) + (1 - 2x + 2y)U_5. \tag{6.74b}$$

The energy norm is

$$a_h(U, U) = \iint_{T_1} (U_x^{[1]^2} + U_y^{[1]^2})\, dx\, dy + \iint_{T_2} (U_x^{[2]^2} + U_y^{[2]^2})\, dx\, dy$$

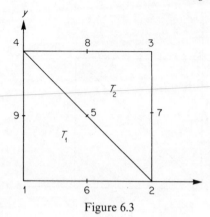

Figure 6.3

and minimizing with respect to U_5 leads to

$$U_5 = 1.$$

Substituting this value back into (6.74a) and (6.74b) leads to

$$U^{[1]}(x, y) = U^{[2]}(x, y) = x + y$$

and so

$$U_h = u$$

throughout the patch. In fact, this is true for any $u \in P_1$ and for any patch of elements, so this nonconforming element passes the patch test.

As a further illustration of the patch test consider the fourth-order problem consisting of the biharmonic equation in a square region with function and normal derivative given round the boundary. The square is divided up in a standard manner into right-angled triangular elements of equal areas, and again we consider the unit square patch comprising two triangular elements (Figure 6.3). For the biharmonic equation the energy contains second-order derivatives and so $r = 2$. In each triangle we consider a quadratic which matches the function values at the vertices of the triangle and the normal derivatives at the midpoints of the sides, i.e. the *Morley triangle* referred to in Chapter 3. This leads to

$$U^{[1]}(x, y) = (1 - x - y + 2xy)U_1 + \tfrac{1}{2}(x + y + x^2 - 2xy - y^2)U_2$$

$$+ \tfrac{1}{2}(x + y - x^2 - 2xy + y^2)U_4 + y(1 - y)\left(\frac{\partial U}{\partial y}\right)_6$$

$$- \frac{1}{\sqrt{2}}(x + y - x^2 - 2xy - y^2)\left(\frac{\partial U}{\partial n}\right)_5 + x(1 - x)\left(\frac{\partial U}{\partial x}\right)_9$$

and

$$U^{[2]}(x, y) = \tfrac{1}{2}(3x - y + y^2 - x^2 - 2xy)U_2 + (1 - x - y + 2xy)U_3$$

$$+ \tfrac{1}{2}(3y - x - y^2 + x^2 - 2xy)U_4 + x(x - 1)\left(\frac{\partial U}{\partial x}\right)_7$$

$$+ y(y - 1)\left(\frac{\partial U}{\partial y}\right)_8$$

$$+ \frac{1}{\sqrt{2}}(-2 + 3x + 3y - x^2 - 2xy - y^2)\left(\frac{\partial U}{\partial n}\right)_5$$

in the triangles T_1 and T_2 respectively, where n is the outward normal to triangle T_1 and the inward normal to triangle T_2. Once again the overall interpolant is not in general continuous across the interface of the two triangles and so the elements are nonconforming. In a fourth-order problem the elements are still nonconforming when the interpolant is continuous across the interface but the normal derivative to the interface is not.

Consider as an example the test solution within the patch to be

$$u = x^2 + y^2$$

leading to the boundary values

$$u_1 = 0 \quad u_2 = u_4 = 1 \quad u_3 = 2$$

$$\left(\frac{\partial u}{\partial y}\right)_6 = \left(\frac{\partial u}{\partial x}\right)_9 = 0 \quad \left(\frac{\partial u}{\partial x}\right)_7 = \left(\frac{\partial u}{\partial y}\right)_8 = 2$$

and hence to the modified interpolants

$$U^{[1]}(x, y) = (x + y - 2xy) - \frac{1}{\sqrt{2}}(x + y)(1 - x - y)\left(\frac{\partial u}{\partial n}\right)_5 \qquad (6.75a)$$

and

$$U^{[2]}(x, y) = (2 - 3x - 3y + 2x^2 + 2xy + 2y^2)$$

$$+ \frac{1}{\sqrt{2}}(x + y - 2)(1 - x - y)\left(\frac{\partial u}{\partial n}\right)_5. \qquad (6.75b)$$

The energy norm is

$$a_h(U, U) = \iint_{T_1} (U_{xx}^{[1]^2} + 2U_{xy}^{[1]^2} + U_{yy}^{[1]^2})\,dx\,dy$$

$$+ \iint_{T_2} (U_{xx}^{[2]^2} + 2U_{xy}^{[2]^2} + U_{yy}^{[2]^2})\,dx\,dy$$

and minimizing with respect to the parameter $(\partial u/\partial n)_5$ leads to

$$\left(\frac{\partial u}{\partial n}\right)_5 = \sqrt{2}.$$

Substituting this value into (6.75a) and (6.75b) leads to

$$U^{[1]}(x, y) = U^{[2]}(x, y) = x^2 + y^2$$

and so

$$U_h = u$$

throughout the patch. In fact this is true for any $u \in P_2$ and for any patch of elements, so the Morley triangle passes the patch test.

Although it is pleasing to verify mathematically that nonconforming elements do or do not pass the patch test, it is enough from a practical point of view to verify it experimentally. The elements are accepted as passing the patch test if the solution reproduces the exact answer, within round-off.

Also for second-order problems, it follows from the earlier convergence analysis (6.88)–(6.71) that the patch test is passed if

$$\int_E (V_h^{[2]} - V_h^{[2]})\,ds = 0 \tag{6.76}$$

where E is any interior edge of the mesh and V_h is any nonconforming function such that $V_h^{[1]}$ and $V_h^{[2]}$ are the limit values as E is approached from opposite sides (Brown, 1975).

Exercise 6.22 Verify that (6.70) is satisfied for the piecewise linear nonconforming elements described above.

Exercise 6.23 Assuming (6.70) apply the Bramble–Hilbert lemma to (6.69) and hence prove (6.71).

Exercise 6.24 Repeat the patch test calculation above for the elements shown in Figure 6.4 and show that

$$U_5 = \frac{1 - \alpha - \alpha^2 + 2\alpha^3}{1 - 2\alpha + 2\alpha^2}.$$

Hence prove that the elements pass the patch test only when $\alpha = \frac{1}{2}$.

Exercise 6.25 Show that the triangular element consisting of the full quadratic, interpolating the values of the function at the vertices and the midpoints of the sides, does not satisfy the patch test for the fourth-order problem described above.

6.5 Convergence of Semi-discrete Galerkin Approximations

This section contains a brief description of one of the convergence estimates developed by Thomée and Wahlbin (1975) and Wheeler (1973) for the model

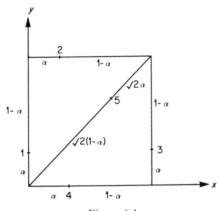

Figure 6.4

problem

$$\frac{\partial u}{\partial t} = \frac{\partial^2 u}{\partial x^2} + \frac{\partial^2 u}{\partial y^2} \quad ((x, y, t) \in R \times (t_0, t_1])$$

subject to the initial condition

$$u(x, y, t_0) = u_0(x, y) \quad ((x, y) \in R)$$

and the boundary condition

$$u(x, y, t) = 0 \quad ((x, y, t) \in \partial R \times (t_0, t_1]).$$

The restriction to two dimensions is not significant.

It is assumed that Theorem 6.5 is valid and there exists $k \geq 1$ such that for any $u \in \mathcal{H}^{(k+1)}(R)$, the error introduced by interpolating u by $\tilde{u} \in K_N$ is bounded as

$$\|u - \tilde{u}\|_{r,R} \leq Ch^{k+1-r}\|u\|_{k+1,R} \quad (r \leq k). \tag{6.77}$$

If it is assumed that the variational crimes outlined in Chapter 5 are avoided, the semi-discrete Galerkin approximation satisfies

$$\left(\frac{\partial U}{\partial t}, V\right) + a(U, V) = 0 \quad (\text{for all } V \in \mathring{K}_N). \tag{6.78}$$

If $u \in \mathcal{H}^{(k+1)}(R) \times C^1[t_0, t_1]$, it follows from (6.77) that the *projection* $W \in \mathring{K}_N \times C^1[t_0, t_1]$ which, for $t \in [t_0, t_1]$, satisfies

$$a(W, V) = a(u, V) \quad (\text{for all } V \in \mathring{K}_N) \tag{6.79}$$

also satisfies

$$\|u - W\|_{r,R} \leq Ch^{k+1-r}\|u\|_{k+1,R} \quad (t \in [t_0, t_1]). \tag{6.80}$$

It follows from (6.78) and (6.79) that

$$\left(\frac{\partial U}{\partial t} - \frac{\partial W}{\partial t}, V\right) + a(U - W, V) = \left(\frac{\partial u}{\partial t} - \frac{\partial W}{\partial t}, V\right)$$

for any $t \in [t_0, t_1]$; with $V = U_t - W_t$ this becomes

$$\| U_t - W_t \|_{0,R}^2 + \frac{1}{2}\frac{d}{dt}a(U - W, U - W) = (u_t - W_t, U_t - W_t).$$

Applying the Schwarz inequality to the right-hand side leads to

$$\| U_t - W_t \|_{0,R}^2 + \frac{1}{2}\frac{d}{dt}a(U - W, U - W) \leqslant \| u_t - W_t \|_{0,R}^2 + \| U_t - W_t \|_{0,R}^2.$$

It follows from (6.80) that

$$\| u_t - W_t \|_{0,R} \leqslant Ch^{k+1} \| u \|_{k+1,R}$$

hence

$$\frac{1}{2}\frac{d}{dt}a(U - W, U - W) \leqslant Ch^{2k+2} \| u \|_{k+1,R}^2.$$

and so

$$a(U - W, U - W) \leqslant Ch^{2k+2} \quad (C = C(u)). \tag{6.81}$$

As the bilinear form a is $\mathscr{H}^{(1)}$ elliptic, combining (6.80) and (6.81) leads to

$$\| u - U \|_{1,R} \leqslant Ch^k.$$

This is a bound on the error in the continuous form of Galerkin approximation. If (6.78) is solved by a step-by-step procedure an additional source of errors would have to be included in the analysis. Error estimates for step-by-step solutions of Galerkin methods have been derived by several authors; additional references can be found in, for example, Dendy (1975) and de Boor (1974).

7

Developments and Applications

7.1 Introduction: Product Approximation

In Chapter 4 we introduced product approximation in which a nonlinear function of the approximation solution was replaced by a *linear* combination of nonlinear functions of *nodal values*. The purpose of this form of approximation is to simplify the construction of the element matrices, but in certain nonlinear equations the product approximation can lead to a more accurate solution than the consistent Galerkin approximation.

Consider, for example, the nonlinear hyperbolic equation

$$u_t + uu_x = 0. \tag{7.1}$$

This can be put in *conservation form* as

$$u_t + \frac{\partial}{\partial x}\left(\frac{u^2}{2}\right) = 0. \tag{7.2}$$

A semi-discrete Galerkin approximation leads to

$$\left(\sum_i \dot{U}_i \varphi_i, \varphi_j\right) - \tfrac{1}{2}\left(\left(\sum_i U_i \varphi_i\right)^2, \varphi_j'\right) = 0 \tag{7.3}$$

which with a piecewise linear approximation becomes

$$\frac{1}{6}(\dot{U}_{j-1} + 4\dot{U}_j + \dot{U}_{j+1}) + \frac{1}{6h}(U_{j+1} + U_j + U_{j-1})(U_{j+1} - U_{j-1}) = 0. \tag{7.4}$$

If product approximation is used on the nonlinear term the system (7.3) is replaced by

$$\left(\sum_i \dot{U}_i \varphi_i, \varphi_j\right) - \tfrac{1}{2}\left(\sum_i U_i^2 \varphi_i, \varphi_j'\right) = 0. \tag{7.5}$$

Note that the second term has been integrated by parts to avoid differentiating

215

the nonlinear term. With piecewise linears, (7.5) becomes

$$\tfrac{1}{6}(\dot{U}_{j+1} + 4\dot{U}_j + \dot{U}_{j-1}) + \frac{1}{4h}(U^2_{j+1} - U^2_{j-1}) = 0. \tag{7.6}$$

The semi-discrete Galerkin approximation to the *linear* equation

$$u_t + u_x = 0 \tag{7.7}$$

has an error that is $O(h^4)$. By comparing the Galerkin approximation to (7.7) with the product approximation to (7.2), we see that the $O(h^4)$ accuracy is retained by (7.6), whereas direct Taylor expansion shows that (7.4) is only $O(h^2)$ accurate (Christie *et al.*, 1981).

7.2 Reaction–Diffusion

In areas such as biology, chemistry and physiology, systems involving several reacting species with diffusion as a transport mechanism are common phenomena. The mathematical model of a process involving reaction and diffusion usually consists of a system of second-order nonlinear partial differential equations of the form

$$\frac{\partial \mathbf{u}}{\partial t} = D\nabla^2 \mathbf{u} + \sum_{j=1}^{m} M_j(\mathbf{x}, \mathbf{u}) \frac{\partial \mathbf{u}}{\partial x_j} + \mathbf{f}(\mathbf{u}) \tag{7.8}$$

where $\mathbf{u}(x, t)$ is an $\mathbb{R}^m$–valued function of $\mathbf{x} \in \mathbb{R}^m$, and $t \in \mathbb{R}^+$. The diffusion matrix D has nonnegative constant entries and is usually diagonal, the coefficients $M_j(j = 1, 2, \ldots, m)$ are continuous matrix-valued functions, the nonlinear function $\mathbf{f}$ describes the reaction of the system, and ∇^2 is the Laplace operator. If (7.8) holds in a region $R \times [t \geqslant 0]$ where R is in m-space with boundary ∂R, we also have the initial condition

$$\mathbf{u}(\mathbf{x}, 0) = \mathbf{u}_0(\mathbf{x}) \quad \mathbf{x} \in R. \tag{7.9}$$

For initial boundary-value problems, we have the additional condition

$$P\frac{\partial \mathbf{u}}{\partial n} + Q\mathbf{u} = \mathbf{a} \quad (\mathbf{x}, t) \in \partial R \times (0, \infty) \tag{7.10}$$

on the boundary, where P and Q are matrix-valued functions of $\mathbf{x}$ and t, $\mathbf{a}$ depends on $\mathbf{x}$ and t, and $\partial/\partial n$ denotes differentiation along a normal to the boundary. The solutions of (7.8)–(7.10) include 'travelling waves' given by

$$\mathbf{u}(\mathbf{x}, t) = \mathbf{U}(\mathbf{x}^T \boldsymbol{v} - ct)$$

where $\boldsymbol{v}$ is a unit vector and c is a scalar velocity. Particular cases of these waves are:

(i) wave fronts $\mathbf{U}(-\infty)$ and $\mathbf{U}(+\infty)$ exist and are unequal,

(ii) pulses $U(\pm\infty)$ exist and are equal; U not constant,

(iii) wave trains (U periodic),

where $U(-\infty)$ and $U(+\infty)$ in (i) and (ii) are zeros of $f(u)$. Wave fronts arise from scalar models like Fisher's equation and pulses and wave trains from the Fitzhugh–Nagumo and Hodgkin–Huxley systems. More complicated wave-like phenomena such as target patterns and spiral waves arise in models of the Belousov–Zhabotinskii reaction. The type of travelling wave obtained in a problem depends on the particular form of $f(u)$, on the initial and boundary conditions (7.9) and (7.10), and usually on the spatial domain being unbounded.

7.2.1 Fisher's equation ($m = n = 1$)

This is the simplest model of reaction–diffusion and takes the form

$$\frac{\partial u}{\partial t} = \frac{\partial^2 u}{\partial x^2} + f(u), \tag{7.11}$$

We consider two types of nonlinearity, viz.

$$\text{(I)} \quad f(u) = u(1 - u) \tag{7.12a}$$

$$\text{(II)} \quad f(u) = u(1 - u)(u - a) \quad 0 < a \leqslant \tfrac{1}{2}. \tag{7.12b}$$

Analytic results depend on replacing (7.11) by the ordinary differential equation

$$U'' - cU' + f(U) = 0 \tag{7.13}$$

where

$$U(\xi) = U(x + ct) = u(x, t).$$

Here c (> 0) is the speed of a wave travelling to the left and a prime denotes differentiation with respect to ξ ($-\infty < \xi < +\infty$). Equation (7.13) can be written as the first-order system

$$V' - cV + f(U) = 0$$

$$U' - V \qquad\quad = 0$$

and so the rest states are given by

$$f(U) = 0.$$

For $f(u) = u(1 - u)$, the rest states $u = 0, 1$ are unstable and stable respectively, and there is a one-parameter family of solutions (Ablowitz *et al.*, 1973) given by

$$U(\xi) = \{1 - r\exp(-\xi/\sqrt{6})\}^{-2} \tag{7.14}$$

where $\xi = x + 5t/\sqrt{6}$, and r is a parameter. It follows from (7.14) that $U(+\infty) = 1$ and $U(-\infty) = 0$. Also the solutions with $r > 0$ blow up, when

$x \to -\infty$, U behaves like

$$r^{-2} \exp(2\xi/\sqrt{6}). \tag{7.15}$$

McKean (1970) has also shown that wave solutions with asymptotic speed $(t \to \infty)$

$$c = \beta + \frac{1}{\beta} \qquad 0 < \beta \leqslant 1 \tag{7.16}$$

can be obtained provided the initial data $u_0(x)$ is a nonnegative continuous monotonic function such that

$$\lim_{x \to -\infty} u_0(x) = \lambda e^{\beta x} \quad \lambda > 0 \tag{7.17}$$

and

$$u_0(+\infty) = 1.$$

It is worth pointing out that for (7.15) and (7.17) to agree,

$$\beta = 2/\sqrt{6}$$

which on substitution into (7.16) gives $c = 5/\sqrt{6}$, a value of the velocity which coincides with that for which Ablowitz *et al.* (1973) obtained the exact solution (7.14).

   Finally, on the theoretical side, Kolmogorov *et al.* (1937) have shown that if the initial data is chosen such that

$$u_0(x) = \begin{cases} 0 & x < 0 \\ 1 & x > 0 \end{cases} \tag{7.18}$$

then the solution approaches a travelling wave of speed $c = 2$.

   In problems of this type, it is essential that the numerical analyst collects all available theoretical information about the problem before embarking on numerical experiments. Here the main goal is to establish that *numerical* solutions of the *partial* differential equation (7.11) together with appropriate initial and boundary data evolve asymptotically in time to the *theoretical* solutions of the *ordinary* differential equation (7.11), on which all analytical results are based. The numerical method proposed is the Petrov–Galerkin procedure similar to that described in Section 4.3 where we approximate u by U, with

$$U(x, t) = \sum_{i=1}^{N} U_i(t)\varphi_i(x).$$

The weak form of (7.11) is

$$\left(\frac{\partial u}{\partial t}, \psi_j\right) + a(u, \psi_j) = (f(u), \psi_j) + \left\langle \frac{\partial u}{\partial x}, \psi_j \right\rangle \quad j = 1, 1, 2, \ldots, N \tag{7.19}$$

where

$$a(u, v) = \left(\frac{\partial u}{\partial x}, \frac{\partial v}{\partial x} \right)$$

and the test functions $\psi_j, j = 1, 2, \ldots, N$, are not necessarily the same as the trial functions φ_j. The boundary term $\langle \partial u/\partial x, \psi_j \rangle$ is zero except possibly when $j = 1$ or N. Product approximation is used to deal with the nonlinear term. The discretization in time of (7.19) is carried out using the predictor–corrector pair (cf. (5.32a) and (5.32b))

$$\left(\frac{1}{\Delta t}(W^{(n+1)} - U^{(n)}), \psi_j \right) + a\left(\frac{W^{(n+1)} + U^{(n)}}{2}, \psi_j \right)$$

$$= (f(U^{(n)}), \psi_j) + \left\langle \frac{\partial U^{(n)}}{\partial x}, \psi_j \right\rangle \tag{7.20a}$$

and

$$\left(\frac{1}{\Delta t}(U^{(n+1)} - U^{(n)}), \psi_j \right) + a\left(\frac{U^{(n+1)} + U^{(n)}}{2}, \psi_j \right)$$

$$= \left(f\left(\frac{W^{(n+1)} + U^{(n)}}{2} \right), \psi_j \right) + \left\langle \frac{\partial U^{(n)}}{\partial x}, \psi_j \right\rangle \tag{7.20b}$$

where $t = n\Delta t, n = 0, 1, \ldots, \Delta t$ is constant, and $W^{(n+1)}$ is the predicted estimate of $U^{(n+1)}$.

Numerical experiments using (7.20) with $f(u) = u(1 - u)$ were carried out over the domain $A \leqslant x \leqslant B$ where A and B are chosen far enough apart to handle any given Dirichlet boundary conditions at $x = \pm \infty$. The numerical results which fully substantiate the theory are as follows:

(i) The initial condition is (7.18), $A = -100$, and $B = +100$. The numerical solution with piecewise linear test and trial functions tends to a travelling front with asymptotic speed 2.

(ii) The initial condition is

$$U_0(x) = \begin{cases} \frac{1}{2}e^{\beta x} & x \leqslant 0 \\ 1 - \frac{1}{2}e^{-\beta x} & x > 0 \end{cases}$$

which satisfies the boundary condition (7.17). The numerical results using piecewise linear trial functions and piecewise B-spline test functions are shown in Table 7.1, and confirm McKean's asymptotic speed (7.16).

(iii) The initial condition is taken from (7.14) for the parameter values $r = -1$, -3 and -10. The trial and test functions, the ranges of x and t, and the grid spacings are all identical with those used in (7.18) and the front propagates with an asymptotic speed $\sim 5/\sqrt{6}$.

Table 7.1

β	c (numerical)	$\beta + 1/\beta$
0.5	2.449	2.500
0.95	2.003	2.003
1.00	1.998	2.000
1.40	1.995	
5.00	1.994	

$-200 \leqslant x \leqslant 100,\ 0 \leqslant t \leqslant 50,$
$h = k = 0.5$

(iv) Finally, the boundary conditions are

$$u = 0 \qquad |x| = \tfrac{1}{2}l, \quad t \geqslant 0$$

where the space domain is $-\tfrac{1}{2}l \leqslant x \leqslant +\tfrac{1}{2}l$ and the initial data is

$$u_0(x) = 0.1\left(1 - \frac{4}{l^2}x^2\right).$$

The numerical solution grows when $l > \pi$ and dies out when $l \leqslant \pi$. This is in agreement with theory for the spatial distribution of the spruce budworm.

For $f(u) = u(1-u)(u-a)$, $0 < a \leqslant \tfrac{1}{2}$, the rest states $u = 0, 1$ are stable and $u = a$ is unstable. When $0 < a \leqslant \tfrac{1}{2}$, equation (7.11) has the Huxley solution

$$U(\xi) = (1 + \exp(-\xi/\sqrt{2}))^{-1} \tag{7.21}$$

where $\xi = x + ct$ with $c = \sqrt{2}(\tfrac{1}{2} - a)$. Again it follows from (7.21) that $U(+\infty) = 1$ and $U(-\infty) = 0$. Numerical calculations are carried out using (7.20) along with piecewise linear trial and test functions. We choose $a = \tfrac{1}{4}$ and impose homogeneous Neumann conditions at the boundaries. When the initial condition is taken from (7.21), the wave front propagates with a speed $\sim \sqrt{2}/4$. For a rectangular pulse initial condition, the pulse either dies away or grows to unity according to whether the initial height of the pulse $\leqslant$ or $> \tfrac{1}{4}$. In the latter case we get two travelling fronts, one in each direction, obeying the Huxley formula with regard to shape and velocity of propagation.

7.2.2 The Fitzhugh–Nagumo (FN) system

This is a more advanced model of reaction–diffusion and governs the condition of electrical impulses in a nerve axon. It is given by

$$\frac{\partial u}{\partial t} = \frac{\partial^2 u}{\partial x^2} + u(1-u)(u-a) - v \qquad 0 < a < \tfrac{1}{2} \tag{7.22a}$$

$$\frac{\partial v}{\partial t} = b(u - dv) \qquad b \geqslant 0, d \geqslant 0 \tag{7.22b}$$

Here $u(x, t)$ is the electrical potential across the axon and $v(x, t)$ is a recovery function. The parameters a, b, and d represent the amount of Novocaine in the system, the reciprocal of the time scale of the recovery process (to the rest state), and a measure of the recovery respectively. The system (7.22) is a simplified model of the Hodgkin–Huxley system (1952), and reduces to Fisher's equation when $b = 0$.

Introducing the variable

$$\xi = x - ct \qquad c \text{ constant}$$

(7.22) becomes the first-order system

$$W' = U(U - a)(U - 1) + V - cW \tag{7.23}$$

$$V' = -\frac{b}{c} U + \frac{bd}{c} V$$

$$U' = W$$

where $u(x, t) = U(\xi)$, $v(x, t) = V(\xi)$, and a prime denotes differentiation with respect to ξ. Note that unlike Fisher's equation we have, for convenience, waves travelling to the *right* with velocity c. If we put $U' = V' = W' = 0$ in (7.23), we obtain the *rest state* solutions

$$U = 0, \tfrac{1}{2}[(1 + a) \pm \{(1 - a)^2 - 4/d\}^{1/2}]. \tag{7.24}$$

From (7.24) it follows that if $d < \{2/(1 - a)\}^2$, the only real rest state is $U = V = W = 0$. Since biological considerations (Hastings, 1981) require the even more restrictive condition

$$d < 3(1 - a + a^2)^{-1}$$

the remainder of this section will be devoted to the single rest state $U = V = W = 0$. The principal weakness in the replacement of (7.22) by (7.23) is that the initial and boundary conditions of the original problem are not retained. Nevertheless, much useful information can be obtained from the first-order system (7.23) and the numerical studies to follow based on the original system (7.22) are influenced significantly by analytical results based on (7.23) together with experimental findings.

Before describing our numerical method for solving (7.22), we list alternative boundary conditions for a semi-infinite nerve axon $0 \leqslant x \leqslant L$, *where L can be as large as required*. At $x = 0$, we have one of

$$u(0, t) = \begin{cases} I & 0 \leqslant t \leqslant T \\ 0 & t > T \end{cases} \tag{7.25a}$$

$$u(0, t) = I \quad t \geqslant 0 \tag{7.25b}$$

$$u(0, t) = \begin{cases} I & n(T_1 + T_2) \leqslant t \leqslant n(T_1 + T_2) + T_1 \\ 0 & n(T_1 + T_2) + T_1 < t < (n+1)(T_1 + T_2) \end{cases} \quad n = 0, 1, \ldots \tag{7.25c}$$

$$\frac{\partial u}{\partial x}(0, t) = -\tfrac{1}{2}I \quad t \geqslant 0. \tag{7.25d}$$

In all four cases, I is constant, and above some positive threshold value. At $x = L$, the boundary condition is taken to be one of the following:

$$u = 0 \tag{7.26a}$$

$$\frac{\partial u}{\partial x} = 0 \tag{7.26b}$$

$$\frac{\partial u}{\partial t} + c \frac{\partial u}{\partial x} = 0 \qquad c \text{ a chosen constant.} \tag{7.26c}$$

The initial condition, unless stated otherwise, is

$$u(x, 0) = 0 \qquad 0 \leqslant x \leqslant L. \tag{7.27}$$

The numerical scheme proposed for the solution of (7.22) subject to the initial condition (7.27) and a pair of boundary conditions, one from each of (7.25) and (7.26), is the Galerkin method in which only one equation (7.22a) is converted to the weak form

$$\left(\frac{\partial u}{\partial t}, w \right) = - a(u, w) + (f, w) - (v, w) + \left\langle \frac{\partial u}{\partial x}, w \right\rangle \quad \forall w \in \mathcal{H}^{(1)} \tag{7.28a}$$

and

$$\frac{\partial v}{\partial t} = b(u - dv) \tag{7.28b}$$

where

$$a(u, w) = \left(\frac{\partial u}{\partial x}, \frac{\partial w}{\partial x} \right)$$

and

$$\left\langle \frac{\partial u}{\partial x}, w \right\rangle \text{ is a boundary term.}$$

The semi-discrete approximations are

$$U(x, t) = \sum_{i=1}^{N} U_i(t) \varphi_i(x) \tag{7.29}$$

and

$$V(x, t) = \sum_{i=1}^{N} V_i(t)\varphi_i(x)$$

respectively where $\varphi_i(x)$, $i = 1, 2, \ldots, N$ are suitable trial functions and $U_i(t)$, $V_i(t)$, $i = 1, 2, \ldots, N$ are time-dependent coefficients. Substitution of (7.29) into the first equation (7.28a) leads to

$$\left(\frac{\partial u}{\partial t}, \psi_j\right) + a(U, \psi_j) = (f, \psi_j) - (V, \psi_j) + \left\langle \frac{\partial u}{\partial x}, \psi_j \right\rangle \quad j = 1, 2, \ldots, N. \quad (7.30)$$

Discretization in time of (7.30) is now carried out according to the predictor–corrector pair

$$\left(\frac{1}{\Delta t}(W^{(n+1)} - U^{(n)}), \psi_j\right) + a\left(\frac{W^{(n+1)} + U^{(n)}}{2}, \psi_j\right)$$

$$= (f(U^{(n)}), \psi_j) - (V^{(n+1)}, \psi_j) + \left\langle \frac{\partial U^{(n)}}{\partial x}, \psi_j \right\rangle$$

$$\left(\frac{1}{\Delta t}(U^{(n+1)} - U^{(n)}), \psi_j\right) + a\left(\frac{U^{(n+1)} + U^{(n)}}{2}, \psi_j\right)$$

$$= \left(f\left(\frac{U^{(n)} + W^{(n+1)}}{2}\right), \psi_j\right) - (V^{(n+1)}, \psi_j) + \left\langle \frac{\partial U^{(n)}}{\partial x}, \psi_j \right\rangle \quad (7.31)$$

and the second equation in (7.28) leads to

$$\frac{1}{\Delta t}(V^{(n+1)} - V^{(n)}) = \tfrac{1}{2}b\{(U^{(n+1)} + U^{(n)}) - d(V^{(n+1)} + V^{(n)})\} \quad (7.32)$$

where $t = n\Delta t$, $n = 0, 1, 2, \ldots$, Δt is constant, and $W^{(n+1)}$ is the predicted estimate of $U^{(n+1)}$.

Equation (7.32) is used to eliminate $V^{(n+1)}$ from (7.31). Unless otherwise stated, the trial functions $\varphi_i(x)$, $i = 1, 2, \ldots, N$, and the test functions $\psi_j(x)$, $j = 1, 2, \ldots, N$, are taken to be piecewise linear functions. It should be noted that the predictor–corrector pair (7.31) avoids the solution of a nonlinear system of equations at each time step.

Numerical calculations using (7.31) and (7.32) are now carried out with appropriate initial and boundary conditions for a range of the parameters a, b, d, and I. The most interesting (and relevant) boundary condition at $x = 0$ is given by (7.25d) and so this is used along with (7.26b) at $x = L$ and the initial condition (7.27). The interested reader is invited to carry out calculations for the other boundary conditions at $x = 0$. The results of the numerical experiments are shown in Table 7.1. In each of the four sets of experiments one parameter only is allowed to vary and the number of pulses (N) generated is shown. In each

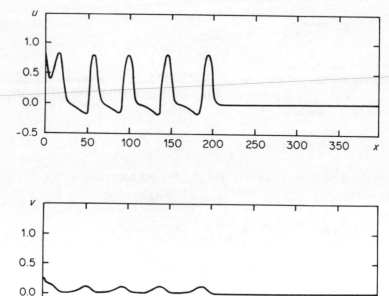

Figure 7.1 Repetitive behaviour of the potential (u) and the corresponding behaviour of the recovery variable (v)

numerical experiment $j = \Delta t = 0.5$ and the length of each calculation is 1000 time steps. The reference point in the four-parameter space is

$$a = 0.139 \quad b = 0.008 \quad d = 2.54 \quad I = 0.6$$

values suggested by Rinzel (1980) and used earlier by Fitzhugh (1968).

A typical case of a train of pulses (repetitive firing) is shown in Figure 7.1 (from Mitchell and Manaranjan, 1982); more details of the numerical experiments leading to Table 7.1 can be found in Manaranjan (1981).

Although more accuracy in the above numerical calculations can be obtained, for example by using cubic spline test functions along with piecewise linear trial functions, it is the authors' experience that linear basis functions usually suffice. The ease with which good numerical solutions are obtained for reaction–diffusion problems is due to two main factors:

(1) the parabolic nature of the differential operator (first-order derivatives in space are not significant), and
(2) the rest states are incorporated without error in the discretized system.

7.3 Solitons

In this section, we turn our attention to travelling wave problems where the interaction between *nonlinearity* and *dispersion* results in solitary waves called *solitons*. A nonlinear wave given by $u_t + uu_x = 0$ has velocity u, and when points of large u overtake points of small u, a shock wave forms which ultimately breaks. However, when we add dispersion in the form of u_{xxx} to the nonlinear wave equation, we get the Korteweg–de-Vries (KdV) equation

$$u_t + uu_x + u_{xxx} = 0 \qquad (7.33)$$

the solution of which is a solitary wave or soliton, owing to the fine balancing of the nonlinear and dispersive effects. A soliton has the remarkable property that in a collision with another soliton it preserves its original shape and speed although a phase shifts exists after the collision.

Equations other than the KdV equation which give rise to soliton solutions are the Sine–Gordon (SG) equation

$$u_{tt} - u_{xx} + \sin u = 0 \qquad (7.34a)$$

and the nonlinear Schrödinger (NLS) equation

$$iu_t + u_{xx} + |u|^2 u = 0. \qquad (7.34b)$$

General surveys of theoretical and numerical developments concerning solitons can be found in papers and books by Bullough and Coudrey (1980), Lamb (1980), Whitham (1974), Eilbeck (1978), Scott *et al.* (1973), and the references therein.

7.3.1 The KdV equation

The KdV equation is a simplified model of the full Euler equation for long waves. The waves, which propagate in one direction only, have long wavelength and small amplitude, but not independently so.

Numerical studies so far have involved finite difference methods and we now look at Galerkin studies for the numerical solution of the pure *initial value* problem involving (7.33). We approximate $u(x, t)$ by

$$U(x, t) = \sum_i U_i(t)\varphi_i(x)$$

where the trial functions $\{\varphi_i(x)\}$ are piecewise linear functions. The unknowns $U_i(t)$ are determined from the system of ordinary differential equations

$$(U_t + UU_x + U_{xxx}, \psi_j) = 0 \quad \text{for all } j$$

where $\{\psi_j(x)\}$ are test functions to be selected. This leads to

$$(U_t + UU_x, \psi_j) - \left(U_{xx}, \frac{\partial \psi_j}{\partial x}\right) = 0 \quad \text{for all } j \qquad (7.35)$$

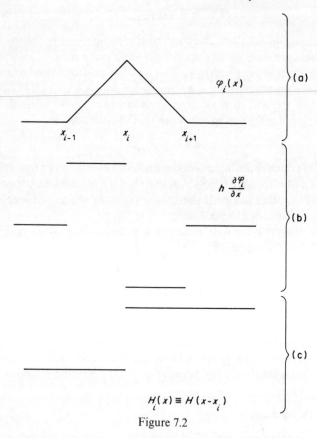

$$H(x) \equiv H(x - x_i)$$

Figure 7.2

after integrating by parts the inner product involving the third derivative and neglecting boundary terms. Since the trial functions are the piecewise linear functions, we get (see Figure 7.2)

$$h\frac{\partial \varphi_i}{\partial x} = H_{i-1} - 2H_i + H_{i+1} \qquad H_i \equiv H(x - x_i)$$

and

$$h^2\frac{\partial^2 \varphi_i}{\partial x^2} = \delta_{i-1} - 2\delta_i + \delta_{i+1} \qquad \delta_i \equiv \delta(x - x_i)$$

where H and δ are the Heaviside and Delta functions respectively, and so from (7.36) $\partial \psi_j / \partial x$ is required to be single-valued at $x = (i - 1)h$, ih, $(i + 1)h$. Possible choices for ψ_j are piecewise quadratic splines (support $3h$) or piecewise cubic splines (support $4h$) centred at $x = jh$.

Exercise 7.1 Using product approximation (see Section 7.1) along with piecewise linear trial functions and piecewise cubic spline test functions centred at the nodes obtain the semi-discrete method

$$\frac{1}{120}(U_{j+2} + 26U_{j+1} + 66U_j + 26U_{j-1} + U_{j-2})$$

$$+ \frac{1}{48h}(U_{j+2}^2 + 10U_{j+1}^2 - 10U_{j-1}^2 - U_{j-2}^2)$$

$$+ \frac{1}{2h^3}(U_{j+2} - 2U_{j+1} + 2U_{j-1} - U_{j-2}) = 0$$

for the KdV equation. Show that the method has a local truncation error at the grid points of $O(h^4)$ (Sanz Serna and Christie, 1981).

From the point of view of ease of computation it would be nice if we could use piecewise linear test functions (linear splines with support $2h$). This of course is not possible (see (7.35)) unless we shift say a distance αh ($0 < \alpha \leqslant \frac{1}{2}$) to the left (or right) of the node, thus producing a four point difference formula which is the lowest order permissible for the KdV equation. This is our motivation for considering the test functions $\{\psi_j(x)\}$ to be $\{\varphi_{j-\alpha}(x)\}$, $0 < \alpha \leqslant \frac{1}{2}$, and so the KdV equation, using product approximation, leads to

$$(\sum_i \{U_i(t)\varphi_i(x) + \tfrac{1}{2}U_i^2(t)\varphi_i'(x)\}, \varphi_{j-\alpha}(x)) - (\sum_i U_i(t)\varphi_i''(x), \varphi_{j-\alpha}'(x)) = 0 \quad \forall j$$

$$(7.36)$$

where $0 \leqslant \alpha \leqslant \frac{1}{2}$. Now *for each value of j*, there are four values of i, viz. $i = j + k$ ($k = -2, -1, 0, +1$) for which the integrals

$$\int_{-\infty}^{+\infty} \varphi_i \varphi_{j-\alpha} \, dx, \quad \int_{-\infty}^{+\infty} \varphi_i' \varphi_{j-\alpha} \, dx, \quad \int_{-\infty}^{+\infty} \varphi_i'', \varphi_{j-\alpha}' \, dx$$

do not vanish. If we label these integrals ha_k, b_k, and c_k/h^2 respectively, it is an easy matter to show that

$$a_{-2} = (\tfrac{1}{6})\alpha^3, \; a_{-1} = (\tfrac{1}{6})(1 + 3\alpha + 3\alpha^2 - 3\alpha^3)$$
$$a_0 = (\tfrac{1}{6})(4 - 6\alpha^2 + 3\alpha^3), \; a_1 = (\tfrac{1}{6})(1 - \alpha)^3$$
$$b_{-2} = -(\tfrac{1}{2})\alpha^2, \; b_{-1} = -(\tfrac{1}{2})(1 - \alpha)(1 + 3\alpha)$$
$$b_0 = -(\tfrac{1}{2})\alpha(-4 + 3\alpha), \; b_1 = (\tfrac{1}{2})(1 - \alpha)^2$$
$$c_{-2} = 1, \; c_{-1} = -3, \; c_0 = 3, \; c_1 = -1.$$

Finally, a Crank–Nicolson discretization in time is used to solve (7.36). This method for solving the KdV equation has been analysed by Schoombie (1982). He proves, and confirmation is given by numerical results, that the best value of the shift parameter is $\alpha = \frac{1}{2}$. This gives the maximum convergence rate of $O(h^2)$ in the discrete $\mathscr{L}_2$ norm and leads to unconditional stability when a Crank–Nicolson time discretization is used.

If the piecewise *quadratic* spline (support $3h$), test function normally centred at $x = jh$, is shifted a distance $\alpha h\,(0 < \alpha \leqslant \frac{1}{2})$ to the left, we get a four point difference scheme in space if $\alpha = \frac{1}{2}$ and a five point scheme if $0 \leqslant \alpha < \frac{1}{2}$. The rate of convergence is $O(h^2)$ unless $\alpha = 1/(2\sqrt{3})$ when the optimum rate is $O(h^3)$.

Numerical studies are now presented for the single soliton solution of the KdV equation

$$u(x, t) = 3c \, \text{sech}^2 (k(x - ct) + d)$$

with $c = 0.2$, $k = (c/4\varepsilon)^{1/2}$ where ε (a coefficient placed in front of the third

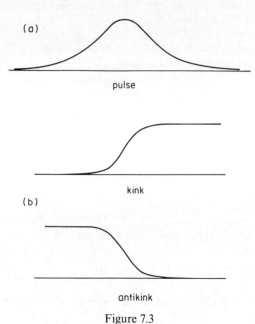

(a)

pulse

kink

(b)

antikink

Figure 7.3

Table 7.2 Error $\times 10^3$ (in discrete $\mathcal{L}_2$ norm). h (step length in space) $= \Delta t$ (step length in time) $= 0.2$

t	Linear $\alpha = \frac{1}{2}$	Quadratic $\alpha = \frac{1}{2}\sqrt{3}$	Cubic $\alpha = 0$
1	8.8	6.0	6.0
5	37	29	28
10	70	58	55
15	102	89	82
20	134	121	108
25	167	156	135

derivative term) $= 0.000484$, $d = -k$, and the initial condition

$$u(x, 0) = 3c \operatorname{sech}^2 (kx + d).$$

This type of soliton is in the form of a pulse and is shown in Figure 7.3(a). The three methods tested have *spline* test functions as shown in Table 7.2 where the numerical results are presented. Further results are presented in Schoombie (1982) and Mitchell and Schoombie (1981).

7.3.2 The Sine–Gordon equation

Another equation which gives rise to solitary waves (solitons) is the Sine–Gordon (SG) equation

$$u_{xx} - u_{tt} = \sin u. \tag{7.37}$$

This time because of the presence of the *second*-order hyperbolic operator, the waves can propagate in *two* directions. A nontrivial solution of (7.37) is a solitary wave in the form of a front given by

$$u = 4 \tan^{-1} \left[\pm \exp \left(\pm \frac{x - ct}{(1 - c^2)^{1/2}} + \alpha \right) \right]$$

where c is the velocity of the wave and α is an arbitrary constant. There are four possible sign combinations with similar signs leading to *kinks* and opposite signs to *antikinks* (see Figure 7.3b).

We now look at the Galerkin method for the numerical solution of the pure *initial value* problem involving (7.37). We approximate U and $\sin U$ by the expansions

$$U(x, t) = \sum_i U_i(t) \varphi_i(x)$$

$$\sin (U(x, t)) = \sum_i \sin (U_i(t)) \varphi_i(x)$$

the latter formula being another example of product approximation. So the Galerkin approximation satisfies

$$\left(\frac{\partial^2 U}{\partial t^2}, \varphi_j \right) + a(U, \varphi_j) + (\sin U, \varphi_j) = 0 \quad \forall j. \tag{7.38}$$

where again

$$a(u, v) = \left(\frac{\partial u}{\partial x}, \frac{\partial v}{\partial x} \right).$$

There is no boundary term since we impose homogeneous Neumann conditions here and after discretization in time (7.38) gives

$$\frac{1}{(\Delta t)^2}(U^{(n+1)} - 2U^{(n)} + U^{(n-1)}, \varphi_j) + a\left(\frac{U^{(n+1)} + U^{(n-1)}}{2}, \varphi_j\right) + (\sin U, \varphi_j) = 0$$

$$(7.39)$$

where $t = n\Delta t$, and the inclusion of $\frac{1}{2}(U^{(n+1)} + U^{(n-1)})$ instead of $U^{(n)}$ is a similar mechanism to that employed by Du Fort and Frankel (1953) to obtain a stable explicit method for a parabolic equation, and by Ewing (1979) for second-order hyperbolic equations.

Further numerical treatment of the SG equation and the more general Klein–Gordon equation can be found in Manaranjan (1981) and Mitchell and Schoombie (1981).

Exercise 7.2 By choosing the basis functions $\{\varphi_j\}$ to be piecewise linear functions, show that (7.39) becomes

$$\frac{h^2}{(\Delta t)^2} M(\mathbf{U}^{(n+1)} - 2\mathbf{U}^{(n)} + \mathbf{U}^{(n-1)}) + \frac{1}{2}S(\mathbf{U}^{(n+1)} + \mathbf{U}^{(n-1)}) + h^2 M \sin(\mathbf{U}^{(n)}) = 0 \quad (7.40)$$

where

$$M = \frac{1}{6}\begin{bmatrix} 2 & 1 & & & \\ 1 & 4 & 1 & & \\ & \ddots & \ddots & \ddots & \\ & & 1 & 4 & 1 \\ & & & 1 & 2 \end{bmatrix} \quad \text{and} \quad S = \begin{bmatrix} 1 & -1 & & & \\ -1 & 2 & -1 & & \\ & \ddots & \ddots & \ddots & \\ & & -1 & 2 & -1 \\ & & & -1 & 1 \end{bmatrix}$$

Exercise 7.3 After replacing M of (7.40) by M_1 (mass lumping), where

$$M_1 = \begin{bmatrix} \frac{1}{2} & & & & \\ & 1 & & & \\ & & \ddots & & \\ & & & 1 & \\ & & & & \frac{1}{2} \end{bmatrix}$$

use Taylor series expansions to show that (7.40) has fourth-order local accuracy. Also using a von Neumann stability analysis, show that (7.40), with M replaced by M_1, has unconditional stability in the linearized sense.

7.4 Conduction–Convection

The convection operator, which involves first-order space derivatives, is nonsymmetric and application of the Galerkin finite element method to problems involving this operator often produces spurious oscillations in the results (see

Section 4.5). This is similar to the behaviour of central difference type finite difference methods, where the oscillations are removed by using 'upwind' difference schemes at the expense of reduced accuracy due to excessive numerical diffusion.

7.4.1 Time-dependent problems

An appropriate time-dependent model of conduction–convection in one space dimension is

$$\frac{\partial u}{\partial t} = \varepsilon \frac{\partial^2 u}{\partial x^2} - q \frac{\partial u}{\partial x} \quad (\varepsilon, q > 0) \tag{7.41}$$

where the initial and boundary conditions are taken to be

$$u(x, 0) = u_0(x) \quad x \in [0, 1]$$

and

$$u(0, t) = u(1, t) = 0 \quad t \geq 0$$

respectively. Applying the semi-discrete Galerkin procedure of Section 5.3, generalized to allow different trial and test functions, we get

$$\sum_{i=1}^{n-1} [(\varphi_i, \psi_j)\dot{U}_i + \{\varepsilon(\varphi_i', \psi_j') + q(\varphi_i', \psi_j)\}U_i] = 0 \quad j = 1, 2, \ldots, n-1$$

where a prime denotes differentiation with respect to x, a dot denotes differentiation with respect to t, and $\varphi(x)$ and $\psi(x)$ are the trial and test functions respectively. Evaluation of (7.41) using piecewise linear trial functions and test functions leads to

$$h^2 M\dot{U} = \varepsilon[(1 + \alpha \mathrm{Pe})A - \mathrm{Pe}B]U \tag{7.42}$$

where the matrices M, A, and B are given by

$$M = \frac{1}{6}\begin{bmatrix} 4 & 1 - \frac{3}{2}\alpha & & & \\ 1 + \frac{3}{2}\alpha & 4 & 1 - \frac{3}{2}\alpha & & \\ & \cdot & \cdot & \cdot & \\ & & \cdot & \cdot & -\frac{3}{2}\alpha \\ & & & 1 + \frac{3}{2}\alpha & 4 \end{bmatrix} \quad A = \begin{bmatrix} -2 & 1 & & \\ 1 & -2 & 1 & \\ & \cdot & \cdot & \cdot & \\ & & & \cdot & \cdot & 1 \\ & & & 1 & -2 \end{bmatrix}$$

and

$$B = \begin{bmatrix} 0 & 1 & & \\ -1 & 0 & 1 & \\ & \cdot & \cdot & \cdot \\ & & \cdot & \cdot & 1 \\ & & & -1 & 0 \end{bmatrix}$$

and $U = [U_1, U_2, \ldots, U_{n-1}]^\mathrm{T}$, with $\mathrm{Pe} = qh/2\varepsilon$ the cell Peclet number.

The overall *stiffness* of the matrix system (7.42) depends on the location of the eigenvalues of the matrix $M^{-1}[(1 + \alpha Pe)A - Pe]$ in the complex plane. This information is essential in choosing a suitable time-stepping algorithm (Lambert, 1973). The interested reader is directed to Mitchell and Griffiths (1979) where an analysis of the stiffness of the system (7.42) is carried out for a range of numbers for both the upwinded linear and quadratic Petrov–Galerkin methods.

7.4.2 The nonlinear case: Burgers' equation

In applications, nonlinear problems are most important, and we now turn to the nonlinear example of conduction–convection given by Burgers' equation

$$\frac{\partial u}{\partial t} = \varepsilon \frac{\partial^2 u}{\partial x^2} - u \frac{\partial u}{\partial x} \quad (0 < x < 1, t > 0) \tag{7.43}$$

with the initial condition

$$u(x, 0) = u_0(x) \quad (0 < x < 1)$$

and the homogeneous boundary conditions

$$u(0, t) = u(1, t) = 0 \quad (t \geq 0)$$

where $\varepsilon (> 0)$ is the coefficient of kinematic viscosity. The mathematical properties of (7.43), which arises in model studies of turbulence and shock wave theory, have been studied by Cole (1951). It is known that for small values of ε the solution develops steep fronts and conventional numerical methods are likely to produce results which include large nonphysical oscillations unless the element size is unrealistically small.

We now apply the Petrov–Galerkin method outlined in the previous section to Burgers' equation resulting in

$$\sum_{i=1}^{n-1} [(\varphi_i, \psi_j)\dot{U}_i + \varepsilon(\varphi'_i, \psi'_j)U_i] + \left(\sum_{i=1}^{n-1} U_i \varphi_i \sum_{i=1}^{n-1} U_i \varphi'_i, \psi_j \right) = 0$$

$$j = 1, 2, \ldots, (n-1). \tag{7.44}$$

With piecewise linear trial functions and the perturbed test functions, equation (7.44) becomes

$$\begin{aligned}
\tfrac{1}{12}h^2 [(2 + 3\alpha)\dot{U}_{j-1} &+ 8\dot{U}_j + (2 - 3\alpha)\dot{U}_{j+1}] \\
&= \varepsilon[U_{j-1} - 2U_j + U_{j+1}] \\
&\quad - \tfrac{1}{12}h[2(U_{j-1} + U_j + U_{j+1})(U_{j+1} - U_{j-1}) \\
&\quad\quad - 3\alpha(U_{j-1}^2 - 2U_j^2 + U_{j+1}^2)] \quad (j = 1, 2, \ldots, n-1)
\end{aligned}$$

where α is an upwinding parameter at our disposal.

Example

Using (7.45) solve Burgers' equation with $\varepsilon = 10^{-4}$, the initial condition $u(x, 0) = \sin \pi x$, and two values of the upwinding parameter (1) $\alpha = 0$, and (2) $\alpha = \alpha_j = \coth(hU_j^{(n)}/2\varepsilon) - 2\varepsilon/(hU_j^{(n)})$ (the latter value is obtained from the steady linear case).

The equations (7.45) together with $U_0 = U_n = 0$ are assembled into a matrix system such as (7.42) where the right-hand side now depends on U in a nonlinear manner. The advancement of the solution in time is carried out by the Crank–Nicolson method with a time increment $\Delta t = 0.001$ and a space increment $h = 1/18$. The set of nonlinear algebraic equations at each time level is solved by a Newton–Raphson method. The smallness of the time increment ensures that any numerical error is due to the discretization in space. The numerical results obtained at $t = 1.0$ are shown in Table 7.3. The accurate solution quoted was computed by a Petrov–Galerkin method with fully upwinded cubic functions (Christie and Mitchell, 1978) and a very small value of h.

The main features of the table are:

(i) The overall poorness of the results.
(ii) The significant improvement due to upwinding.

Table 7.3

| Node number | Accurate solution | Linear elements | |
		No upwinding	Upwinding
0	0.0	0.0	0.0
1	0.0422	0.108	0.0421
2	0.0843	−0.113	0.0842
3	0.1263	0.405	0.1263
4	0.1684	−0.443	0.1682
5	0.2103	0.713	0.2103
6	0.2522	−0.936	0.2518
7	0.2939	1.093	0.2942
8	0.3355	−1.401	0.3344
9	0.3769	1.607	0.3784
10	0.4182	−0.369	0.4148
11	0.4592	1.306	0.4650
12	0.5000	−0.384	0.4879
13	0.5404	1.419	0.5627
14	0.5806	−0.216	0.5357
15	0.6203	1.399	0.7041
16	0.6596	−0.115	0.4861
17	0.6983	1.466	1.0053
18	0.0	0.0	0.0

The principal conclusion is that it is essential to reduce the grid size in the vicinity of steep gradients of the solution. In fact, a strategy based on upwinding combined with grid reduction in selected parts of the range may be the way to proceed in nonlinear problems of conduction–convection type. This point is taken up again in a later section of the present chapter.

Numerical results comparing a variety of finite element and finite difference methods for a two-dimensional conduction–convection problem are given by Griffiths and Mitchell (1979). The test problem chosen, where some sort of theoretical solution is available, is due to Raithby (1976). Unfortunately, this problem has 'weak' outflow boundary conditions and so no boundary layer.

7.5 Singular Isoparametric Elements

In Section 3.3.5 it was shown that the nodes of isoparametric elements have to be placed with care to ensure that the Jacobian does not vanish within the element. There are, however, circumstances under which a Jacobian with a zero at a particular point can be an advantage. In this section we outline one application of such isoparametric elements.

If the function $u(x, y)$ satisfies

$$\frac{\partial^2 u}{\partial x^2} + \frac{\partial^2 u}{\partial y^2} = 0$$

in a region R, then *in the neighbourhood of a corner* in the boundary ∂R, it follows that u can be written as

$$u = \sum_{j=1}^{\infty} \gamma_j r^{j/\alpha} \sin\left(\frac{j\theta}{\alpha}\right) \tag{7.45}$$

for some constants $\gamma_j (j = 1, \ldots)$, where $\alpha\pi$ is the angle subtended by R at the corner and (r, θ) are the polar coordinates with the corner as origin. Thus it follows that near a re-entrant corner ($\alpha > 1$), the derivatives of the leading term in (7.45) are unbounded as r tends to zero. The two most common situations are $\alpha = 2$—a region with a slit or crack—and $\alpha = \frac{3}{2}$—a region with a right-angled 'elbow'. The leading terms in the expansion then become proportional to $r^{1/2}$ and $r^{2/3}$ respectively. One of the main reasons for the failure of many standard numerical methods for such problems is the inability of polynomials (in r) to represent such functions sufficiently accurately.

We now give two examples of isoparametric elements that overcome this difficulty, provided that the corner of the region is taken as a vertex of an element in which the other nodes are placed in a special way.

(1) *Quadratic elements* can be used to represent the $r^{1/2}$ behaviour. If $t_4 = \frac{1}{2}(t_1 + t_2)$, $t_5 = \frac{1}{4}(3t_3 + t_2)$, and $t_6 = \frac{1}{4}(3t_3 + t_1)$ (see Figure 3.7), then the

isoparametric transformation becomes

$$t - t_3 = ((t_1 - t_3)p + (t_2 - t_3)q)(p + q) \quad (t = x, y) \tag{7.46}$$

and linear functions of p and q have the necessary $r^{1/2}$ form, where r is the distance from the vertex P_3. For example along $P_1 P_3 (q = 0)$, it follows that

$$r^2 = (x - x_3)^2 + (y - y_3)^2$$
$$= ((x_1 - x_3)^2 + (y_1 - y_3)^2)p^4$$

and so $p \approx r^{1/2}$. The Jacobian of this transformation can be written as

$$2C_{123}(p + q)^2$$

and hence only vanishes at $P_3 (p = q = 0)$.

(2) *Cubic elements* can be used to represent $r^{2/3}$ behaviour in the neighbourhood of the node P_3. When the nodes are positioned correctly, the transformation becomes

$$t - t_3 = ((t_1 - t_3)p + (t_2 - t_3)q)(p + q)^2 \quad (t = x, y).$$

Thus, by analogy with the quadratic case, linear functions of p and q behave as $r^{1/3}$ and hence quadratic functions have the necessary $r^{2/3}$ behaviour. In this cubic case, the Jacobian of the transformation is

$$3C_{123}(p + q)^4$$

which only vanishes when $p = q = 0$.

Biquadratic elements can be used with the midside nodes at the $\frac{1}{2}$-point positions, but such elements are usually discounted as crack-tip elements as the $r^{1/2}$ behaviour is only exhibited along grid lines. In $\frac{1}{4}$-point triangles, it follows from (7.46) that lines

$$\alpha(x - x_3) = \beta(y - y_3)$$

in the physical plane are mapped onto lines

$$\alpha p = \beta q$$

in the reference coordinates and along such lines $r^2 \approx R^4$ where

$$R^2 = p^2 + q^2$$

and

$$r^2 = (x - x_3)^2 + (y - y_3)^2. \tag{7.47}$$

In order to achieve sufficient accuracy in the neighbourhood of the singularity it is usually necessary to refine the grid in some way. If a band of singular elements is clustered around the singularity then, as the grid is refined, the region covered by such elements is reduced and the effect of the elements on the solution is reduced. In the limit as the size of the singular elements tends to zero, so does the beneficial effect of the $\frac{1}{4}$-points. To overcome this drawback it is necessary to include *transition* elements in the region between the singular elements and the regular elements a fixed distance from the singularity.

The transition elements are arranged so that, as the grid is refined, the region covered by singular elements and transition elements remains constant. Within the transition elements it is necessary that the same $r^{1/2}$ behaviour occurs but with the origin just outside the element. Thus in place of (7.47) we have

$$r^2 = (x - X)^2 + (y - Y)^2$$

and

$$R = (p - P)^2 + (q - Q)^2$$

where (X, Y) is the fixed singularity in (x, y)-coordinates and (P, Q) are the local reference coordinates of the image point.

The most straightforward construction of transition elements is to view the fixed region to be covered by a combination of singular and transition elements as a single-macro-element. If this macro-element is defined by means of a $\frac{1}{4}$-point transformation and then subdivided by means of a quasi-regular subdivision of the reference macro-element, in the physical (x, y)-plane we have a small $\frac{1}{4}$-point element adjacent to the singularity together with a graded grid of transition elements.

Consider the example, illustrated by Figure 7.4, of a mesh of unit right-angled triangles away from a singularity at the origin and where the triangle adjacent to the singularity is replaced by a $\frac{1}{4}$-point macro-element denoted by M.

The reference macro-element that is the image of M under the $\frac{1}{4}$-point

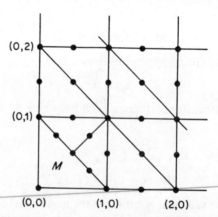

Figure 7.4 Elements outside macro-element M involve linear transformations

transformation is subdivided as illustrated in Figure 7.5. The transformation can be written as

$$x = p(p + q)$$

$$y = q(p + q)$$

and the coordinates of the nodes in M are given in Table 7.4 and Figure 7.6.

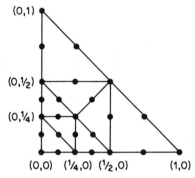

Figure 7.5 Subdivision of reference macro-element

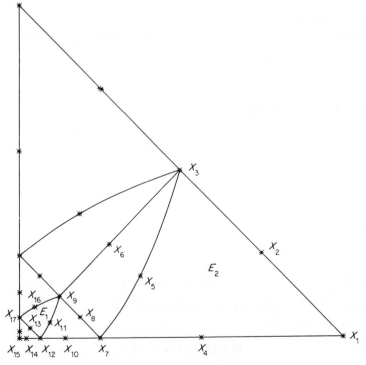

Figure 7.6 The graded grid within the macro-element in (x, y)-space

Table 7.4 Coordinates of nodes X_N in macro-element M

N	p	q	x	y
1	1	0	1	0
2	$\frac{3}{4}$	$\frac{1}{4}$	$\frac{3}{4}$	$\frac{1}{4}$
3	$\frac{1}{2}$	$\frac{1}{2}$	$\frac{1}{2}$	$\frac{1}{2}$
4	$\frac{3}{4}$	0	$\frac{9}{16}$	0
5	$\frac{1}{2}$	$\frac{1}{4}$	$\frac{3}{8}$	$\frac{3}{16}$
6	$\frac{3}{8}$	$\frac{3}{8}$	$\frac{9}{32}$	$\frac{9}{32}$
7	$\frac{1}{2}$	0	$\frac{1}{4}$	0
8	$\frac{3}{8}$	$\frac{1}{8}$	$\frac{3}{16}$	$\frac{1}{16}$
9	$\frac{1}{4}$	$\frac{1}{4}$	$\frac{1}{8}$	$\frac{1}{8}$
10	$\frac{3}{8}$	0	$\frac{9}{64}$	0
11	$\frac{1}{4}$	$\frac{1}{8}$	$\frac{3}{32}$	$\frac{3}{64}$
12	$\frac{1}{4}$	0	$\frac{1}{16}$	0
13	$\frac{1}{8}$	$\frac{1}{8}$	$\frac{1}{32}$	$\frac{1}{32}$
14	$\frac{1}{8}$	0	$\frac{1}{64}$	0
15	0	0	0	0
16	0	$\frac{3}{8}$	0	$\frac{9}{64}$
17	$\frac{1}{8}$	$\frac{1}{4}$	$\frac{3}{64}$	$\frac{3}{62}$

Clearly, from the list of coordinates, it follows that the small element adjacent to the singularity is a $\frac{1}{4}$-point element with the local transformation

$$x = \tfrac{1}{16}p(p + q)$$
$$y = \tfrac{1}{16}q(p + q).$$

Consider the transition element E_1, with nodes (from Table 7.4) X_9, X_{11}, X_{12}, X_{13}, X_{16}, and X_{17}. To simplify the algebra, map these onto the reference element using a quadratic transformation T, such that

$$T_1 : (0, 0) \mapsto X_9$$
$$T_1 : (1, 0) \mapsto X_{17}$$
$$T_1 : (0, 1) \mapsto X_{12}.$$

Then the transformation becomes

$$x = \tfrac{1}{16}(1 - p)(2 - p - q)$$
$$y = \tfrac{1}{16}(1 - q)(2 - p - q). \tag{7.48}$$

In terms of the local reference coordinates for E_1, the singular point is at $p = q = 1$ and from (7.48) with

$$R^2 = (1 - p)^2 + (1 - q)^2 \tag{7.49}$$

it follows that a line segment $\alpha x = \beta y$ is mapped onto a line segment

$\beta(1 - p) = \alpha(1 - q)$ along which

$$R^2 \approx r^2.$$

Similarly in the element E_2 if the quadratic transformation T_2 is such that

$$T_2:(1,0) \mapsto X_1$$
$$T_2:(0,1) \mapsto X_3$$
$$T_2:(0,0) \mapsto X_7$$

then the transformation becomes

$$x = \tfrac{1}{4}(1 + p)(1 + p + q)$$
$$y = \tfrac{1}{4}q(1 + p + q). \tag{7.50}$$

In terms of the local reference coordinates, the singularity is at $p = -1, q = 0$ and if

$$R^2 = (1 + p)^2 + q^2$$

the image of $\alpha x = \beta y$ is $\alpha(1 + p) = \beta q$ and again $R^4 \approx r^2$.

Further details of this formulation of singular elements and transition elements, together with the results of numerical experiments, can be found in Wait (1977, 1979a).

7.6 First-order Hyperbolic Equations

In this section we consider a few of the problems associated with solving the two-dimensional advection equation

$$u_t + au_x + bu_y = 0 \quad \begin{cases} t > 0 \\ (x, y) \in R \end{cases} \tag{7.51}$$

with a fixed uniform mesh. In order to formulate a well posed problem with a first-order hyperbolic equation it is only possible to specify the *inflow boundary condition*, that is to specify the flow $u(x, y, t)$ only along those sections of the boundary ∂R for which

$$\mathbf{a}^T \mathbf{n} < 0$$

where $\mathbf{a}^T = (a, b)$ and $\mathbf{n}$ is the outward normal. For a stable numerical approximation, any outflow conditions introduced must not affect the flow set up by the inflow boundary condition.

A typical numerical problem is to model the flow in an infinite region by considering only a finite subregion (e.g. flow round an aerofoil) and hence the boundaries in the numerical model do not exist in a physical sense.

There is no dissipative term in (7.51) and hence it follows from the weak formulation

$$(u_t + au_x + bu_y, u) = 0$$

that with a and b constant

$$\frac{d}{dt} \|u\|_0^2 + \mathbf{a}^{\mathrm{T}}\mathbf{n} \int_{\partial R} u^2 \, ds = 0 \tag{7.52}$$

where the *energy* $\|u\|_0$ is defined as

$$\|u\|_0^2 = \int_R u^2 \, dx \, dy.$$

The derivation of the boundary terms is similar to the proof of Green's theorem in Section 1.2.5.

It follows from (7.52) that

(i) If $R = \mathbb{R}^2$ then the boundary terms disappear and

$$\|u\|_0^2 = \text{constant}$$

i.e. energy is conserved.

(ii) If R is finite then

$$\frac{d}{dt} \|u\|_0^2 = -\mathbf{a}^{\mathrm{T}}\mathbf{n} \int_{\partial R} u^2 \, ds$$

and the energy is increased by waves crossing the inflow boundary and reduced by waves crossing the outflow boundary. In particular if $u = 0$ on the inflow boundary the energy cannot increase.

An accurate numerical solution must attempt to reproduce all the properties of the analytic solution, i.e. no dissipation, no dispersion, no reflection from the outflow boundary, and the waves should travel at the correct speed. With the single linear equation (7.51), since the differential du satisfies

$$du = u_t \, dt + u_x \, dx + u_y \, dy$$

the solution is constant along lines such that

$$\frac{dt}{1} = \frac{dx}{a} = \frac{dy}{b}.$$

These lines are known as *characteristics* and, for this equation, can be written as

$$x - at = \text{constant}$$

$$y - bt = \text{constant}.$$

Thus

$$u(x, y, t) = u(x + \Delta x, y + \Delta y, t + \Delta t)$$

iff

$$\Delta t = \frac{\Delta x}{a} = \frac{\Delta y}{b}$$

and hence if the initial condition is

$$u(x, y, 0) = f(x, y)$$

then the general solution is

$$u(x, y, t) = f(x - at, y - bt).$$

7.6.1 Stability and phase error

Following the style of the von Neumann stability analysis of Section 5.3.3 consider a single component of the solution of (7.51), written as

$$w(t) = e^{i\sigma(x + y)}e^{-ict}. \tag{7.53}$$

The analysis can be modified to cover the general nonsymmetric component with $\sigma(x + y)$ replaced by $\sigma_1 x + \sigma_2 y$ and the overall conclusions remain valid. If (7.53) is substituted in (7.51) it follows that the *phase velocity* $c = c_1$ satisfies

$$c_1 = (a + b)\sigma. \tag{7.54}$$

If w is a component of the semi-discrete piecewise bilinear Galerkin solution defined by

$$(U_t, V) + (aU_x, V) + (bU_y, V) = 0 \tag{7.55}$$

then

$$\left(\frac{\cos(\sigma h) + 2}{3}\right)^2 \dot{w} + (a + b)\left(\frac{i\sin(\sigma h)}{h}\frac{(\cos(\sigma h) + 2)}{3}\right)w = 0$$

and the solution can be written as (7.53) if $c = c_2$ where

$$c_2 = (a + b)\frac{\sin(\sigma h)}{h}\frac{3}{\cos(\sigma h) + 2}. \tag{7.56}$$

If the semi-discrete solution is to move at the same speed as the analytic solution, the phase velocities (7.54) and (7.56) should be equal. It can be seen however that the *phase velocity error*

$$1 - \frac{c_2}{c_1} = 1 - \frac{\sin(\sigma h)}{\sigma h} \cdot \frac{3}{\cos(\sigma h) + 2}$$

depends on σ and indicates that the components of the numerical solution (unlike those of the analytic solution) do not all travel at the same speed. Thus the Galerkin method has introduced numerical *dispersion* that will cause the wave to decompose and break up as it travels. The numerical scheme is *conservative* (*nondissipative*) as the amplification factor

$$\lambda = e^{-ic\Delta t}$$

is such that $|\lambda| = 1$ (or $\mathrm{Re}(ic) = 0$).

If the semi-discrete equations (7.55) are replaced by the CNG equations

$$(U^{(n+1)} - U^{(n)}, V) + \frac{a\Delta t}{2}(U_x^{(n+1)} + U_x^{(n)}, V) + \frac{b\Delta t}{2}\left(U_y^{(n+1)} + U_y^{(n)}, V \right) = 0 \quad (7.57)$$

then the continuous component $w(t)$ in (7.53) is discretized to give

$$w^{(n)} = e^{i\sigma(x+y)} e^{-ic(n\Delta t)}. \quad (7.58)$$

This is a solution of (7.57) if

$$w^{(n+1)} - w^{(n)} + (a+b)\frac{\Delta t}{2}i\frac{\sin(\sigma h)}{\sigma h}\cdot\frac{3}{\cos(\sigma h)+2}(w^{(n+1)} + w^{(n)}) = 0$$

that is

$$w^{(n+1)} = \frac{1 - i\alpha}{1 + i\alpha}w^{(n)} \quad (7.59)$$

where

$$\alpha = (a+b)\frac{\Delta t}{2}\frac{\sin(\sigma h)}{\sigma h}\cdot\frac{3}{\cos(\sigma h)+2}.$$

The phase velocity $c = c_3$ satisfies

$$e^{-ic_3\Delta t} = \frac{1 - i\alpha}{1 + i\alpha}$$

that is

$$c_3\Delta t = \arctan\left(\frac{2\alpha}{1 - \alpha^2}\right)$$

Once again $|e^{-ic\Delta t}| = 1$ and hence there is no dissipation in the numerical scheme. The ratio c_3/c_1 is given in Table 7.5 for various values of σh. It can be seen from Table 7.5 that both the semi-discrete and the CNG solutions lead to approximations that move too slowly (*phase lag*). It is possible to derive other schemes in order to reduce the lag: for example we can replace (7.57) by the alternating direction Galerkin scheme

$$\left(U^{(n+1)} + a\frac{\Delta t}{2} U_x^{(n+1)} + b\frac{\Delta t}{2} U_y^{(n+1)} + ab\left(\frac{\Delta t}{2}\right)^2 U_{xy}^{(n+1)}, V \right)$$

$$= \left(U^{(n)} - a\frac{\Delta t}{2} U_x^{(n)} - b\frac{\Delta t}{2} U_y^{(n)} + ab\left(\frac{\Delta t}{2}\right)^2 U_{xy}^{(n)}, V \right). \qquad (7.60)$$

This can be viewed as a CNG approximation to

$$\left(1 + ab\left(\frac{\Delta t}{2}\right)^2 \frac{\partial^2}{\partial x \, \partial y} \right) u_t + au_x + bu_y = 0$$

where the additional term

$$ab\left(\frac{\Delta t}{2}\right)^2 u_{xyt}$$

modifies the dispersion in the numerical solution. A Fourier component (7.58) is a solution of (7.60) if

$$\left[\frac{\cos(\sigma h) + 2}{3} - ab\left(\frac{\Delta t}{2}\right)^2 \left(\frac{\sin(\sigma h)}{h}\right)^2 \right] (w^{(n+1)} - w^{(n)})$$

$$+ (a+b)\frac{\Delta t}{2} i \frac{\sin(\sigma h)}{h} (w^{(n+1)} + w^{(n)}) = 0$$

which can be written as (7.59) with

$$\alpha = \frac{(a+b)\dfrac{\Delta t}{2}\dfrac{\sin(\sigma h)}{h}}{\left(\dfrac{\cos(\sigma h) + 2}{3}\right)^2 - ab\left(\dfrac{\Delta t}{2}\right)\left(\dfrac{\sin(\sigma h)}{h}\right)^2}.$$

The phase velocity of the ADG solution is denoted by c_4 and can be evaluated for various values of σh as in Table 7.5.

Table 7.5(a) $a = b = 1$, $\Delta t = h/2$, $\sigma h = \pi/N$

N	c_2/c_1	c_3/c_1	c_4/c_1	c_5/c_1
2	0.955	0.819	0.914	1.070
2.5	0.983	0.881	0.954	1.054
3	0.992	0.915	0.971	1.040
3.5	0.996	0.936	0.980	1.031
4	0.998	0.951	0.985	1.024
4.5	0.999	0.961	0.989	1.019
5	0.999	0.968	0.991	1.016

c_2/c_1 is independent of Δt

Table 7.5(b) $a = b = 1$, $\sigma h = \pi/N$

N	c_3/c_1	c_4/c_1	c_5/c_1
$\Delta t = h$			
2	0.626	0.819	1.131
2.5	0.709	0.881	1.142
3	0.768	0.915	1.125
3.5	0.813	0.937	1.105
4	0.843	0.937	1.086
4.5	0.872	0.961	1.071
5	0.892	0.968	1.059
$\Delta t = 0.1h$			
2	0.948	0.953	0.960
2.5	0.978	0.982	0.986
3	0.988	0.991	0.994
3.5	0.993	0.995	0.998
4	0.996	0.998	0.999
4.5	0.997	0.999	0.999

It is possible to construct methods for which the numerical phase velocity is too large (*phase lead*), for example a time discretization can be constructed using the Taylor expansions

$$u\left(T + \frac{\Delta t}{2}\right) \approx u(T) + \frac{\Delta t}{2}\frac{\partial u(T)}{\partial t} + \left(\frac{\Delta t}{2}\right)^2 \frac{1}{2}\frac{\partial^2 u(T)}{\partial t^2}$$

$$\approx u(T + \Delta t) - \frac{\Delta t}{2}\frac{\partial u(T + \Delta t)}{\partial t} + \left(\frac{\Delta t}{2}\right)^2 \frac{1}{2}\frac{\partial^2 u(T + \Delta t)}{\partial t^2} \qquad (7.61)$$

As u is the solution of (7.53), it follows that

$$\frac{\partial u}{\partial t} = -a\frac{\partial u}{\partial x} - b\frac{\partial u}{\partial y}$$

and that

$$\frac{\partial^2 u}{\partial t^2} = \left(a\frac{\partial}{\partial x} + b\frac{\partial}{\partial y}\right)^2 u.$$

Thus (7.61) leads to

$$\left(U^{(n)} - \frac{\Delta t}{2}(aU_x^{(n)} + bU_y^{(n)}) + \left(\frac{\Delta t}{2}\right)^2\left(a\frac{\partial}{\partial x} + b\frac{\partial}{\partial y}\right)^2 U^{(n)}, V\right)$$

$$= \left(U^{(n+1)} + \frac{\Delta t}{2}(aU_x^{(n+1)} + bU_y^{(n+1)}) + \left(\frac{\Delta t}{2}\right)^2\left(a\frac{\partial}{\partial x} + b\frac{\partial}{\partial y}\right)^2 U^{(n+1)}, V\right).$$

Applying integration by parts to the last terms on each side provides the alternative form

$$(U^{(n+1)}, V) + \left(aU_x^{(n+1)} + bU_y^{(n+1)}, \frac{\Delta t}{2} V - \left(\frac{\Delta t}{2} \right)^2 \tfrac{1}{2}(aV_x + bV_y) \right)$$

$$= (U^{(n)}, V) - \left(aU_x^{(n)} + bU_y^{(n)}, \frac{\Delta t}{2} V + \left(\frac{\Delta t}{2} \right)^2 \tfrac{1}{2}(aV_x + bV_y) \right). \tag{7.62}$$

Either form can be viewed as the CNG approximation to the weak form of the equation

$$\left(1 + \left(\frac{\Delta t}{2} \right)^2 \frac{1}{2} \left(a\frac{\partial}{\partial x} + b\frac{\partial}{\partial y} \right)^2 \right) u_t + au_x + bu_y = 0$$

which is (7.53) modified by the addition of another dispersion term.

In the solution of (7.62), a Fourier component (7.58) satisfies

$$\left[\left(\frac{\cos(\sigma h) + 2}{3} \right)^2 \right.$$

$$+ \left(\frac{\Delta t}{2} \right)^2 \left(2(a^2 + b^2) \left(\frac{\sin(\sigma h)}{h} \right)^2 \frac{\cos(\sigma h) + 2}{3} + ab \left(\frac{\sin(\sigma h)}{h} \right)^2 \right) \right] (w^{(n+1)} - w^{(n)})$$

$$+ (a + b)\frac{\Delta t}{2} i \left(\frac{\sin(\sigma h)}{h} \right)(w^{(n+1)} + w^{(n)}) = 0$$

and again this can be rearranged into the form (7.59) to provide values of the phase velocity $c = c_5$. The phase ratio c_5/c_1 is given in Table 7.5 for various values of σh.

As a more severe test of the methods, consider the equation with variable coefficients, i.e.

$$u_t + a(x, y)u_x + b(x, y)u_y = 0$$

where the numerical example to follow concerns a rotating flow

$$u_t + yu_x - xu_y = 0 \quad \begin{cases} t > 0 \\ -\infty < x, y < \infty \end{cases}. \tag{7.63}$$

The *characteristics* are no longer straight lines in (x, y, t)-space and if the initial state is

$$u(x, y, 0) = f(x, y)$$

then, in general,

$$u(x, y, t) = f(x \cos t - y \sin t, x \sin t + y \cos t).$$

The equation (7.63) is formulated as a pure initial value problem (*Cauchy problem*) so that all the boundaries introduced by the numerical method are nonphysical boundaries. The ADG method is applied with natural boundary conditions in the sense that (7.60) is satisfied for all the basis functions V, including those corresponding to the boundary nodes, and that no additional conditions are imposed on the boundary.

The basis functions are in tensor product form (bilinear in Figure 7.7 and biquadratic in Figure 7.8) so that for constant coefficients the discretization can be written as

$$\left(B_x + a\frac{\Delta t}{2}G_x\right) \otimes \left(B_y + b\frac{\Delta t}{2}G_y\right)\mathbf{U}^{(n+1)} = \left(B_x - a\frac{\Delta t}{2}G_x\right) \otimes \left(B_y - b\frac{\Delta t}{2}G_y\right)\mathbf{U}^{(n)}$$

$$(7.64)$$

where B_x, B_y, G_x, and G_y are tridiagonal matrices (cf. Sections 7.4.1 and 7.4.6). As explained in Section 4.5, the solution of system (7.64) is solved in two stages: first row-by-row and then column-by-column, saving both time and storage. In

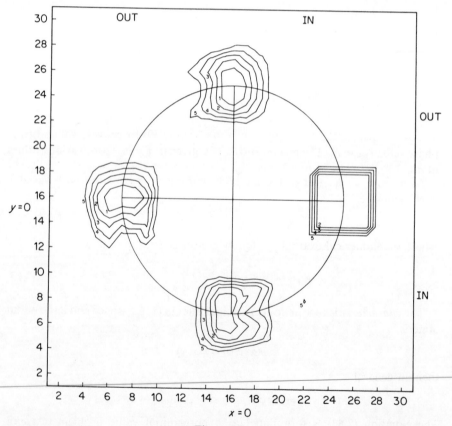

Figure 7.7

the solution of (7.63) the coefficients $a = y$ and $b = -x$ are not constant and hence the tridiagonal matrices

$$\left(B_x + \frac{y\Delta t}{2} G_x \right) \quad \text{and} \quad \left(B_y - \frac{x\Delta t}{2} G_y \right)$$

vary from row to row in the first stage and from column to column in the second stage. It is also assumed in this formulation that the variable coefficients are taken as constant within each element.

The calculations illustrated in Figure 7.7 and 7.8 used $\Delta t = 0.028$ and the same 31×31 grid of nodes; hence $h = 1$ for the bilinears in Figure 7.7 and $h = 2$ for the biquadratics in Figure 7.8. The initial condition was a 4×4 square unit pulse centred on $x = 9$, $y = 0$. As $56\Delta t \approx \frac{1}{2}\pi$ it follows that after 56 time steps the pulse should have rotated (unaltered) through an angle of $\frac{1}{2}\pi$, through π after 112 time steps, and $\frac{3}{2}\pi$ after 168 time steps. The contours are at heights 0.3, 0.5, 0.7, 0.9, and 1.1 and indicate considerable dispersion in the numerical solution. The extent of the dispersion is indicated by the contour surrounding node (22, 8), i.e. $x = 6$, $y =$

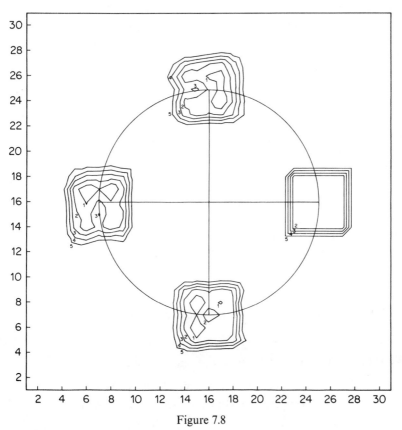

Figure 7.8

− 8 in Figure 7.7. In both solutions there are considerable ripples behind the main pulse as a result of the dispersion. As might be expected from Table 7.5 with the small value of $\Delta t/h$ the phase velocity of the main pulse is very accurate. These results were provided for the authors by M. Parsaei.

Another problem with wave (and advection) equations is the modelling of subgrid scale waves (i.e. $\pi/\sigma h < 2$) and the spurious oscillations in the numerical solution that appear to travel in the wrong direction. A more detailed analysis can be found in Hedstrom (1979) and an alternative approach to the solution of hyperbolic equations in Dupont (1973).

7.7 The Moving Finite Element Method

In the numerical solution of elliptic equations an unequal subdivision of the region is often used in problems where the solution changes significantly in certain parts of the region (cf. Section 7.5). In many important time-dependent problems the parts requiring finer subdivision alter with time and so *adaptive* grids are required. Adaptive procedures, however, are still at the development stage and although their real worth will eventually be in multidimensional computations, we shall illustrate the 'moving finite element method' in *one* space dimension only.

Consider the scalar evolution equation

$$\frac{\partial u}{\partial t} = Au \quad t \geq 0 \tag{7.65}$$

where A is a partial differential operator, involving space derivatives only. In this section we approximate u by a piecewise polynomial U defined on the time-dependent grid

$$\Pi(t); x_0(t) < x_1(t) < \cdots < x_N(t)$$

where $(N + 1)$ is a *fixed* number of nodes.

In most problems x_0 and x_N are fixed and we assume that the boundary values U_0 and U_N are *known constants*. If we write

$$U(x, t) = \sum_{i=0}^{N} U_i(t)\varphi_i \tag{7.66}$$

then if the positions of the nodes are time dependent we have

$$\varphi_i = \varphi_i(x; x_1(t), \ldots, x_{N-1}(t)). \tag{7.67}$$

We restrict our attention to piecewise linear functions

$$\varphi_i(x) = \begin{cases} (x - x_{i-1})/(x_i - x_{i-1}) & x_{i-1} \leq x \leq x_i \\ (x_{i+1} - x)/(x_{i+1} - x_i) & x_i \leq x \leq x_{i+1} \\ 0 & \text{elsewhere.} \end{cases} \tag{7.68}$$

in which case

$$\varphi_i = \varphi_i(x; x_{i-1}(t), x_i(t), x_{i+1}(t)).$$

It follows from (7.66) and (7.67) that

$$\frac{\partial U}{\partial t} = \sum_{i=1}^{N-1} \frac{dU_i}{dt} \varphi_i + \sum_{i=1}^{N-1} \frac{dx_i}{dt} \frac{\partial U}{\partial x_i} \tag{7.69}$$

and if we write $\psi_i \equiv \partial U/\partial x_i$ then for piecewise linear trial functions

$$\psi_i = U_{i-1}\frac{\partial \varphi_{i-1}}{\partial x_i} + U_i\frac{\partial \varphi_i}{\partial x_i} + U_{i+1}\frac{\partial \varphi_{i+1}}{\partial x_i}$$

$$= \begin{cases} U_{i-1}\dfrac{\partial \varphi_{i-1}}{\partial x_i} + U_i\dfrac{\partial \varphi_i}{\partial x_i} & x_{i-1} \leqslant x \leqslant x_i \\[2mm] U_i\dfrac{\partial \varphi_i}{\partial x_i} + U_{i+1}\dfrac{\partial \varphi_{i+1}}{\partial x_i} & x_i \leqslant x \leqslant x_{i+1} \end{cases}$$

and it follows from (7.68) that

$$\psi_i(x) = \begin{cases} -m_i\varphi_i(x) & x_{i-1} \leqslant x \leqslant x_i \\ -m_{i+1}\varphi_i(x) & x_i \leqslant x \leqslant x_{i+1} \\ 0 & \text{elsewhere} \end{cases}$$

where $m_i = (U_i - U_{i-1})/(x_i - x_{i-1})$, and m_{i+1} similarly.

The weak formulation of a differential equation

$$\frac{\partial u}{\partial t} = Au$$

on a moving grid therefore leads to the Galerkin equations

$$\sum_i (\varphi_j, \varphi_i)\dot{U}_i + \sum_i (\varphi_j, \psi_i)\dot{x}_i - (\varphi_j, AU) = 0$$

$$\sum_i (\psi_j, \varphi_i)\dot{U}_i + \sum_i (\psi_j, \psi_i)\dot{x}_i - (\psi_j, AU) = 0 \quad (j = 1, 2, \ldots, N-1) \tag{7.70}$$

(Gelinas *et al.*, 1981; Miller and Miller, 1981; Miller, 1981).

The system of equations (7.70) is now ready for solution either by a Gear package or by a time stepping algorithm constructed to deal with a specific example of (7.65). Like many automatic methods, the adaptive procedure based on (7.70) can run into difficulties, e.g. when

(i) $m_i = m_{i+1}$,
(ii) neighbouring grid points cross over each other,

although the latter can be prevented by incorporating a penalty function into the procedure (see Gelinas *et al.*, 1981). A more serious feature of this method is the discontinuous nature of the basis functions $\psi_i(x)$, $i = 1, 2, \ldots, N - 1$, which occur in several of the inner products in (7.70). In cases where (7.65) involves second-order or higher space derivatives, these inner products cease to exist in the usual sense, although Gelinas *et al.* handle second-order space derivatives by means of a limiting process.

A *Petrov–Galerkin method* (Herbst *et al.*, 1982) has been devised in which the test functions are replaced by cubic basis functions

$$S_j(x) = [\varphi_j(x)]^2[3 - 2\varphi_j(x)]$$

and

$$T_j(x) = [\varphi_j(x)]^2[\varphi_j(x) - 1]\left[\frac{d\varphi_j(x)}{dx}\right]^{-1}$$

respectively. This leads to the equations

$$\sum_i (S_j, \varphi_i)\dot{U}_i + \sum_i (S_j, \psi_i)x_i - (S_j, AU) = 0$$

$$\sum_i (T_j, \varphi_i)\dot{U}_i + \sum_i (T_j, \psi_i)\dot{x}_i - (T_j, AU) = 0 \quad (i, j = 1, 2, \ldots, N - 1). \quad (7.71)$$

Several problems are solved using (7.71) in Herbst *et al.* (1982).

7.7.1 Equidistributing principles

Consider a function $u(x)$, $x \in [0, 1]$ and its piecewise linear approximation $U(x)$ with respect to the grid

$$\Pi : 0 = x_0 < x_1 < \cdots < x_N = 1.$$

If $u(x) \in C^2[0, 1]$, then the error

$$e(x) = u(x) - U(x)$$

on the interval $I_i = [x_{i-1}, x_i]$ of length h_i is bounded by the inequality (Shultz, 1969)

$$\|e\|_i \leqslant \tfrac{1}{8}h_i^2\|u''\|_i$$

where $\|.\|_i$ denotes the maximum norm over the interval I_i. The global error in terms of the maximum norm can now be expressed as

$$\|e\|_\infty = \max_{1 \leqslant i \leqslant N} \|e\|_i \leqslant \tfrac{1}{8} \max_{1 \leqslant i \leqslant N} (h_i^2\|u''\|_i)$$

and the question is whether the nodes x_i, $i = 1, 2, \ldots, N - 1$, can be placed in such a way that the quantity $\|e\|_\infty$ is minimized. The answer according to de Boor

(1973) is to place the nodes so that

$$h_i^2 \|u''\|_i = h_{i+1}^2 \|u''\|_{i+1} \quad (i = 1, 2, \ldots, N-1).$$

The latter is an example of an *equidistributing principle*, where the nodes are placed in such a way that some specified quantity is kept the same on each subinterval of the grid. In two-point boundary value problems, Pereyra and Sewell (1975) have used *approximate equidistributing principles* of the form

$$h_i^2 \|u''\|_i + O(h_i^3) = h_{i+1}^2 \|u''\|_{i+1} + O(h_{i+1}^3) \tag{7.72}$$

as node placement strategies and it is shown in Herbst *et al.* (1982) that the Galerkin and Petrov–Galerkin methods described earlier are equivalent to principles of the form (7.72). Despite this, much work remains to be done on the problem of node placement in time-dependent problems, particularly when multidimensional calculations are involved.

References

Ablowitz, M. J., Kamp, D. J., Newell, A. C., and Segur, H. (1973). *Phys. Rev. Lett.*, **30**, 1262.

Abramowitz, M., and Stegun, I. R. (1964). *Handbook of mathematical functions*, Dover, New York.

Adams, R. A. (1975). *Sobolev Spaces*, Academic Press, New York.

Adini, A., and Clough, R. W. (1961). *Analysis of Plate Bending by the Finite Element Method*, Nat. Sci. Found. Rept. G7337, Univ. of California, Berkeley.

Agmon, S. (1965). *Lectures on Elliptic Boundary Value Problems*, Van Nostrand, Princeton.

Ahlberg, J. H., and Ito, T. (1975). *Math. Comp.*, **29**, 761.

Akin, J. E. (1982). *Application and Implementation of finite element methods*, Academic Press, New York.

Anderssen, R. S., and Mitchell, A. R. (1979). *Math. methods Appld. Sci.*, **1**, 3.

Arthurs, A. M. (1980). *Complementary Variational Principles*, 2nd Ed., Clarendon Press, Oxford.

Aubin, J. P. (1972). *Approximation of Elliptic Boundary Value Problems*, Wiley, New York.

Aubin, J. P. (1977). *Applied Abstract Analysis*, Wiley, New York.

Aubin, J. P. (1979). *Applied Functional Analysis*, Wiley, New York.

Aziz, A. K. (Ed.) (1972). *The Mathematical Foundations of the Finite Element Method with Applications to Partial Differential Equations*, Academic Press, New York.

Babuška, I. (1971). *SIAM J. Numer. Anal.*, **8**, 304.

Babuška, I. (1973). *Numer. Math.*, **20**, 179.

Babuška, I., and Aziz, A. K. (1976). *SIAM J. Numer. Anal.*, **13**, 214.

Babuška, I., and Rosenweig, M. B. (1972). *Numer. Math.*, **20**, 1.

Baker, G. A. (1973). *Math. Comp.*, **27**, 229.

Barnhill, R. E., Gregory, J. A., and Whiteman, J. R. (1972). 749–755 in Aziz (1972).

Barnsley, M. F., and Robinson, P. D. (1977). *J. Inst. Math. Applics.*, **20**, 485.

Berger, A. E. (1972). 757–796 in Aziz (1972).

Berger, A. E. (1973). *Numer. Math.*, **21**, 345.

Berger, A. E., Scott, R., and Strang, G. (1972). 295–313 in *Symposia Mathematica X*, Academic Press, London.

Bers, L., John, F. and Schechter, M. (1964). *Partial Differential Equations*, Interscience, New York.

Birkhoff, G., Schultz, M. H., and Varga, R. S. (1968). *Numer. Math.*, **11**, 232.

Birkhoff, G. (1971). *Proc. Nat. Acad. Sci.*, **68**, 1162.

Bond, T. J., Swanell, R. D., Henshell, R. D., and Warburton, G. B. (1973). *J. Strain Anal.*, **8**, 182.

de Boor, C. (1972). *J. Approx. Th.*, **6**, 50.

de Boor, C. (1973). 12 in Watson (1973).

de Boor, C. (Ed.) (1974). *Mathematical Aspects of Finite Elements in Partial Differential Equations*, Academic Press, New York.

de Boor, C., and Swartz, B. (1973). *SIAM J. Numer. Anal.*, **10**, 582.

Bramble, J. H., Dupont, T., and Thomée, V. (1972). *Math. Comp.*, **26**, 869.

Bramble, J. H., and Hilbert, S. R. (1970). *SIAM J. Numer. Anal.*, **7**, 112.

Bramble, J. H., and Nitsche, J. A. (1973). *SIAM J. Numer. Anal.*, **10**, 81.

Bramble, J. H., and Schatz, A. H. (1970). *Comm. P. Appld. Math.*, **23**, 635.

Bramble, J. H., and Schatz, A. H. (1971). *Math. Comp.*, **25**, 1.

Bramble, J. H., and Thomée, V. (1974). *R.A.I.R.O.*, **8** (R-2), 5.

Bramble, J. H., and Zlámal, M. (1970). *Math. Comp.*, **24**, 809.

Brown, J. H. (1975). *Non Conforming Finite Elements and their Applications*, M.Sc. Thesis, Univ. of Dundee.

Brown, N. G., and Wait, R. (1982). *IMA J. Numer. Anal.*, **2**, 481.

Bullough, R. K., and Coudrey, P. J. (1980). The soliton and its history. In *Solitons*, Springer-Verlag, Berlin, pp. 1–64.

Cecchi, M. M., and Cella, A. (1973). 767–768 in *Proc. 4th Canadian Congress on Appld. Mech.*

Chernuka, M. W., Cowper, G. R., Lindberg, G. M., and Olson, M. D. (1972). *Int. J. Num. Meth. Eng.*, **4**, 49.

Christie, I., Griffiths, D. F., Mitchell, A. R., and Sans Serna, J. M. (1981). *I.M.A. J. Numer. Anal.*, **1**, 253.

Christie, I., and Mitchell, A. R. (1978). *Int. J. Numer. methods Engng.*, **12**, 1764.

Ciarlet, P. G. (1973a). 12 in Watson (1973).

Ciarlet, P. G. (1973b). 113–129 in Whiteman (1973).

Ciarlet, P. G. (1978). *The finite element method for elliptic problems*, North-Holland, Amsterdam.

Ciarlet, P. G., and Raviart, P. A. (1972a). *Arch. Rat. Mech. Anal.*, **46**, 177.

Ciarlet, P. G., and Raviart, P. A. (1972b). *Comp. Meth. App. Mech. Eng.*, **1**, 217.

Ciarlet, P. G., and Raviart, P. A. (1972c). 409–474 in Aziz (1972).

Clegg, J. C. (1967). *Calculus of Variations*, Oliver and Boyd, Edinburgh.

Cole, J. D. (1951). *Quart. App. Math.*, **9**, 225.

Collins, W. D. (1977). *Proc. Roy. Soc. Edin.*, **77A**, 273.

Comini, G., del Guidici, S., Lewis, R. W., and Zienkiewicz, O. C. (1974). *Int. J. Num. Meth. Eng.*, **8**, 613.

Courant, R., and Hilbert, D. (1953). *Methods of Mathematical Physics*, Vol. I, Interscience, New York.

Cox, M. (1972). *J. Inst. Math. Applics.*, **10**, 134.

Crouzeix, M., and Raviart, P. A. (1973). *R.A.I.R.O.*, **7**, (R-3), 33.

Dem'janovič, J. K. (1964). *Sov. Math. Dokl.*, **5**, 1452.

Dendy, J. E. (1975). *SIAM J. Numer. Anal.*, **12**, 541.

Dendy, J. E., and Fairweather, G. (1975). *SIAM J. Numer. Anal.*, **12**, 144.

Diaz, J. C. (1977). *SIAM J. Numer. Anal.*, **14**, 844.

Diaz, J. C. (1979). *Math. Comp.* **33**, 77.

Douglas, J., and Dupont, T. (1970). *SIAM J. Numer. Anal.*, **7**, 575.

Douglas, J., and Dupont, T. (1971). 133–244 in Hubbard (1971).

Douglas, J., and Dupont, T. (1973). *Math. Comp.*, **27**, 17.

Douglas, J., and Dupont, T. (1975). *Math. Comp.*, **29**, 360.

Douglas, J., Dupont, T., and Wheeler, M. F. (1974). *R.A.I.R.O.*, **8** (R-2), 61.

Du Fort, E. C., and Frankel, J. P. (1953). *M.T.A.C.*, **7**, 135.

Dupont, T. (1973). *SIAM J. Numer. Anal.*, **10**, 890.

Dupont, T., Fairweather, G., and Johnson, J. P. (1974). *SIAM J. Numer. Anal.*, **11**, 392.

Eilbeck, J. C. (1978). Numerical studies of solitons. In *Solitons and condensed matter physics*, A. R. Bishop and T. Schnuder (Eds.), Springer-Verlag. p. 28.

Ewing, R. E. (1979). *An efficient time-stepping method for nonlinear second order equations*, M.R.C. 1966, University of Wisconsin.

Fairweather, G. (1978). *Finite element Galerkin methods for differential equations*, Marcel Dekker, Basle.

Fairweather, G., and Johnson, J. P. (1975). *Numer. Math.*, **23**, 269.

Fairweather, G., and Saylor, A. V. (1983). *IMA J. Numer. Anal.*, **3**, 173.

Finlayson, B. A. (1972). *The method of weighted residuals*, Academic Press, New York.

Finlayson, B. A., and Scriven, L. E. (1967). *Int. J. Heat Mass. Trans.*, **10**, 799.

Fitzhugh, R. (1968). 1–85 in H.P. Schwann, (Ed.), *Biological Engineering*, McGraw-Hill, New York.

Fix, G. J. (1972). 525–556 in Aziz (1972).

Forsythe, G. E., and Wasow, W. R. (1960). *Finite difference methods for partial differential equations*, interscience, New York.

Fried, I. (1980). *Int. J. Numer. methods Engng.*, **15**, 451.

Gear, C. W. (1971). *Numerical initial value problems in ordinary differential equations*, Prentice-Hall, Englewood Cliffs.

Gelinas, R. J., Doss, S. K., and Miller, K. (1981). *J. Comp. Phys.*, **40**, 202.

George, A., and Liu, J. W. (1981). *Computer solution of large sparse positive definite systems*, Prentice-Hall, Englewood Cliffs.

Gordon, W. J. (1971). *SIAM J. Numer. Anal.*, **8**, 158.

Gordon, W. J., and Wixom, J. A. (1974). *SIAM J. Numer. Anal.*, **11**, 909.

Gourlay, A. R., and Morris, J. Ll. (1980). *SIAM J. Numer. Anal.*, **17**, 641.

Gourlay, A. R., and Morris, J. Ll. (1981). *IMA J. Numer. Anal.*, **1**, 347.

Gresho, P. M., Lee, R. L., and Stullich, T. W. (1978). 3.45–3.63 in C. A. Brebbia, G. F. Pinder and W. G. Gray (Eds.), *Finite elements in water resources* 2, Pentel, Southampton.

Griffiths, D. F., and Mitchell, A. R. (1979). 91–104 in Hughes (1979).

Hastings, S. (1981). *Math. Biol.*, **11**, 105.

Hayes, L. (1981). *Int. J. Numer. methods Engng.*, **16**, 35.

Hedstrom, G. W. (1979). *SIAM J. Numer. Anal.*, **16**, 385.

Herbold, R. J., and Varga, R. S. (1972). *Aeq. Math.*, **7**, 36.

Herbst, B. M., Mitchell, A. R., and Schoombie, S. W. (1982). *Int. J. Numer. methods Engng.*, **18**, 1321.

Herrera, L., and Sewell, M. J. (1978). *J. Inst. Math. Applics.*, **21**, 95.

Hildebrand, F. B. (1965). *Methods of Applied Mathematics*, Prentice Hall, New York.

Hinton, E., and Owen, D. R. J. (1979). *An introduction to finite element computations*, Pineridge Press, Swansea.

Hodgkin, A. L., and Huxley, A. F. (1952). *J. Physiol.*, **117**, 500.

Hopkins, T. R., and Wait, R. (1978). *Int. J. Numer. methods Engng.*, **12**, 1081.

Hubbard, B. (ED.) (1971). *Numerical Solution of Partial Differential Equations.* II, *SYNSPADE 1970*, Academic Press, New York.

Hughes, T. J. R. (Ed.) (1979). *Finite element methods in convection dominated flows*, AMD 34, ASME, New York.

Hughes, T. J. R., and Brooks, A. (1979). 19–36 in Hughes (1979).

Hulme, B. L. (1972). *Math. Comp.*, **26**, 415.

Hutson, V., and Pym, J. S. (1980). *Applications of functional analysis and operator theory*, Academic Press, New York.

Irons, B. M. (1966). *Conf. on use of digital computers in structural engng.*, Newcastle.

Irons, B. M. (1969). *Int. J. Numer. methods Engng.*, **1**, 29.

Irons, B. M. (1970). *Int. J. Numer. methods Engng.*, **2**, 5.

Irons, B. M., and Loikkanen, M. (1983). *Int. J. Numer. methods Engng.*, **19**, 1391.

Irons, B. M., and Razzaque, A. (1972). 557–587 in Aziz (1972).

Jespersen, W. B. (1978). *SIAM J. Numer. Anal.*, **15**, 813.

Jordan, W. B. (1970). *AEC Research and Development report KAPL-M-7112*.

Kolmogorov, A. N., Petrovkii, I. G., and Piskunov, N. S. (1937). *Bjul Moskovskogo Gas Univ.*, **1**, 1–26.

Lamb, G. L. J. (1980). *Elements of soliton theory*, Wiley, New York.

Lambert, J. D. (1973). *Computational Methods in Ordinary Differential Equations*, Wiley, London.

Lascaux, P., and Lesaint, P. (1975). *R.A.I.R.O.*, **9** (R-1), 9.

Laurie, D. P. (1976). *J. Inst. Math. Applics.*, **19**, 119.

Lawson, J. D., and Morris, J. Ll. (1978). *SIAM J. Numer. Anal.*, **15**, 1212.

Linberg, B. (1971). *B.I.T.*, **2**, 29.

Lions, J. L., and Magenes, E. (1972). *Non-Homogeneous Boundary Value Problems and Applications* I, Springer-Verlag, Berlin.

Lucas, T. R., and Reddien, G. W. (1972). *SIAM J. Numer. Anal.*, **9**, 341.

Manaranjan, V. S. (1981). *A finite element method for solving the Klein–Gordon equation*, N.A./48, University of Dundee.

Matthies, H., and Strang, G. (1979). *Int. J. Numer. methods Engng.*, **14**, 1613.

McKean, H. P. (1970). *Advances in Math.*, **4**, 209.

McLeod, R. J. (1978). *J. Inst. Math. Applics.*, **21**, 419.

McLeod, R. J. (1979). *Computers and Math. Applics.*, **5**, 267.

McLeod, R. J., and Mitchell, A. R. (1972). *J. Inst. Math. Applics.*, **10**, 382.

McLeod, R. J., and Mitchell, A. R. (1975). *J. Inst. Math. Applics.*, **16**, 239.

McLeod, R. J., and Mitchell, A. R. (1979). *Computers and Math. Applics.*, **5**, 277.

Miller, K. (1981). *SIAM J. Numer. Anal.*, **18**, 1033.

Miller, K., and Miller, R. (1981). *SIAM J. Numer. Anal.*, **18**, 1019.

Milne, R. D. (1980). *Applied functional analysis: an introductory treatment*, Pitman, London.

Mitchell, A. R. (1976) in Whiteman (1976).

Mitchell, A. R. (1979). *Computers and Math. Applics.*, **5**, 321.

Mitchell, A. R., and Griffiths, D. F. (1977). 90–104 in Watson (1977).

Mitchell, A.R., and Griffiths, D. F. (1979). 19–35 in Whiteman (1979).

Mitchell, A. R., and Griffiths, D. F. (1980). *The finite difference method in partial differential equations*, Wiley, Chichester.

Mitchell, A. R., and Manaranjan, V. S. (1982). 17–37 in Whiteman (1982).

Mitchell, A. R., and Schoombie, S. W. (1981). Finite element studies of solitons. In Hinton, E., Bettess, P., and Whems, R. (Eds.), *Numerical methods for coupled problems*, Pineridge Press, Swansea.

Morse, P. M., and Feshbach, H. (1953). *Methods of Theoretical Physics*, McGraw-Hill, New York.

Natterer, F. (1975). *Numer. Math.*, **25**, 67.

Natterer, F. (1977). *Math. Comp.*, **31**, 457.

Nečas, J. (1967). *Les Methodes Directes en Théorie des Equations Elliptiques*, Academia, Prague.

Nitsche, J. A. (1971). *Abhandt. d. Hamb. Math. Sem.*, **36**, 9.
Nitsche, J. A. (1972). 603–627 in Aziz (1972).
Nitsche, J. A. (1975). In *Proc. 2nd Conference on finite elements, Rennes.*
Nitsche, J. A. (1976). In Galligani and Magenes (Eds.), *Mathematical aspects of finite element methods*, Lecture notes in Math., **606**, Springer-Verlag, Berlin.
Nitsche, J. A. (1979) in Whiteman (1979).
Nitsche, J. A., and Schatz, A. H. (1974). *Math. Comp.*, **28**, 937.
Noble, B. (1973). 143–152 in Whiteman (1973).
Noble, B., and Sewell, M. J. (1972). *J. Inst. Math. Applics.*, **9**, 123.
Oden, J. T. (1972). *Finite Elements of Nonlinear Continua*, McGraw-Hill, New York.
Oden, J. T., and Reddy, J. N. (1976). *An introduction to the mathematical theory of finite elements*, Interscience, New York.
Oganesyan, L. A. (1966). *USSR Comp. Math. and Math. Phys.*, **6**, 116.
Oganesyan, L. A., and Rukhovets, L. A. (1969). *USSR Comp. Math. and Math. Phys.*, **9**, 153.
Peaceman, D. W., and Rachford, H. H. (1955). *J. Soc. Ind. Appl. Math.*, **3**, 28.
Pereyra, V., and Sewell, E. G. (1975). *Numer. Math.*, **23**, 261.
Pian, T. H. H. (1970). *Numerical Solution of Field Problems in Continuum Physics*, SIAM–AMS Proceedings Volume 2.
Powell, M. J. D. (1973). *Conference on Numerical Software*, Loughborough.
Rachford, H. H., and Wheeler, M. F. (1974) in de Boor (1974).
Raithby, G. D. (1976). *Comp. Methods Appl. Mech. Engng.*, **9**, 153.
Rinzel, J. (1980). In Stewart, W. E., Ray, W. H., and Corley, C. C. (Eds.), *Dynamics and modelling of reactive systems*, Academic Press, New York.
Sanz Serna, J. M., and Christie, I. (1981). *J. Comp. Phy.*, **39**, 94.
Schatz, A. H., and Wahlbin, L. (1977). *Math. Comp.*, **31**, 77.
Schatz, A. H., and Wahlbin, L. (1978). *Math. Comp.*, **32**, 73.
Schatz, A. H., and Wahlbin, L. (1979). *Math. Comp.*, **33**, 465.
Schatz, A. H., and Wahlbin, L. (1981). In Vichnevetsky, R., and Stepleman, R. S. (Eds.), *Advances in computer methods for partial differential equations*, I.M.A.C.S., New Brunswick.
Schatz, A. H., and Wahlbin, L. (1982). *Math. Comp.*, **38**, 1.
Schechter, R. S. (1967). *The variational method in engineering*, McGraw-Hill, New York.
Schoombie, S. W. (1982). *IMA J. Numer. Anal.*, **2**, 95.
Schultz, M. H. (1969). *SIAM J. Numer. Anal.*, **7**, 161.
Scott, R. (1975). *SIAM J. Numer. Anal.*, **12**, 404.
Scott, R. (1976). *Math. Comp.*, **30**, 681.
Scott, A. C., Chu, F. Y. F., and McLoughlin, D. W. (1973). *Proc. IEEE*, **61**, 1443.
Sewell, M. J. (1969). *Phil. Trans. Roy. Soc. (London)*, A **265**, 319.
Sewell, M. J., and Noble, B. (1978). *Proc. Roy. Soc. Lond.*, A **361**, 293.
Showalter, R. E. (1977). *Hilbert space methods for partial differential equations*, Pitman, London.
Siemenuich, J. L., and Gladwell, I. (1974). *Numer. Anal. Report* 5, Manchester University.
Smith, I. M. (1982). *Programming the finite element method with applications to geomechanics*, Wiley, Chichester.
Stakgold, I. (1979). *Green's functions and boundary value problems*, Interscience, New York.
Strang, G. (1972). 689–710 in Aziz (1972).
Strang, G., and Berger, A. R. (1971). 199–205 in *Proc. American Math. Soc. Summer Inst. in Partial Diff. Equns.*
Strang, G., and Fix, G. (1973). *An Analysis of the Finite Element Method*, Prentice Hall, New Jersey.

Stummel, F. (1980). *Int. J. Numer. methods Engng.*, **15**, 177.
Thomée, V. (1973). *J. Inst. Math. Applics.*, **11**, 33.
Thomée, V., and Wahlbin, L. (1975). *SIAM J. Numer. Anal.*, **12**, 378.
Treves, F. (1980). *Introduction to pseudodifferential operators and Fourier integral operators*, Plenum Press, New York.
Varga, R. S. (1971). *Functional Analysis and Approximation Theory in Numerical Analysis*, SIAM Publications, Philadelphia.
Vichnevetsky, R. and Bowles, J. B. (1982). *Fourier analysis of numerical approximations of hyperbolic equations*, Studies 5, SIAM, Philadelphia.
Wachspress, E. L. (1971). Conf. on Appl. Num. Anal., Dundee, *Springer-Verlag Lecture Notes in Math.*, **228**, 223.
Wachspress, E. L. (1973a). *J. Inst. Math. Applics.*, **11**, 83.
Wachspress, E. L. (1973b). 177 in Watson (1973).
Wachspress, E L. (1975). *A Rational Finite Element Basis*, Academic Press, New York.
Wait, R. (1977). *J. Inst. Math. Applics.*, **20**, 131.
Wait, R. (1979a). *J. Inst. Math. Applics.*, **24**, 471.
Wait, R. (1979b). *The numerical solution of algebraic equations*, Wiley, Chichester.
Wait, R. (1985a). *Finite Element Algorithms and Approximations*, Wiley, Chichester.
Wait, R. (1985b). *Finite Element Algorithms and Approximations Software*, Wiley, Chichester.
Washizu, K. (1968). *Variational Methods in Elasticity and Plasticity*, Pergamon Press, London.
Watson, G. A. (Ed.) (1973). Conf. Num. Soln. Diff. Equns., Dundee, *Springer-Verlag Lecture Notes in Maths.*, **363**.
Watson, G. A. (Ed.) (1977). Conf. Num. Anal., Dundee, *Springer-Verlag Lecture Notes in Maths.* **630**.
Wheeler, M. F. (1973). *SIAM J. Numer. Anal.*, **10**, 723.
Wheeler, M. F. (1977). *SIAM J. Numer. Anal.*, **14**, 71.
Wheeler, M. F. (1978). *SIAM J. Numer. Anal.*, **15**, 152.
Whiteman, J. R. (Ed.) (1973). *The Mathematics of Finite Elements and Applications* I, Academic Press, New York.
Whiteman, J. R. (Ed.) (1976). *The Mathematics of Finite Elements and Applications* II, Academic Press, New York.
Whiteman, J. R. (Ed.) (1979). *The Mathematics of Finite Elements and Applications* III, Academic Press, New York.
Whiteman, J. R. (Ed.) (1982). *The Mathematics of Finite Elements and Applications* IV, Academic Press, New York.
Whitham, G. B. (1974). *Linear and nonlinear waves*, Wiley, New York.
Wilson, E. L., Taylor, R. L., Doherty, W. P., and Ghaboussi, J. (1971). *Univ. of Illinois Symposium.*
Witsch, K. (1978a). *Numer. Math.*, **30**, 185.
Witsch, K. (1978b). *Numer. Math.*, **30**, 333.
Woodford, G., Mitchell, A. R., and McLeod, R. (1978). *Int. J. Numer. methods Engng.*, **12**, 1587.
Yosida, K. (1971). *Functional analysis*, Springer-Verlag, Berlin.
Zafarullah, A. (1970). *J. Assoc. Comp. Mech.*, **17**, 294.
Zienkiewicz, O. C. (1977). *The finite element method*, 3rd edition, McGraw-Hill, London.
Zlámal, M. (1973). *SIAM J. Numer. Anal.*, **10**, 227.
Zlámal, M. (1974). *SIAM J. Numer. Anal.*, **11**, 347.
Zlámal, M. (1975). *Math. Comp.*, **29**, 350.

Index